Philosophy, R
and the
Spiritual Life

ROYAL INSTITUTE OF PHILOSOPHY SUPPLEMENT: 32

EDITED BY

Michael McGhee

CAMBRIDGE
UNIVERSITY PRESS

Published by the Press Syndicate of the University of Cambridge
The Pitt Building, Trumpington Street, Cambridge, CB2 1RP
40 West 20th Street, New York, NY 10011-4211, USA
10 Stamford Road, Oakleigh, Victoria 3166, Australia

*A catalogue record for this book is available
from the British Library*

ISBN 0 521 42196 9 (paperback)

Library of Congress Cataloguing in Publication Data

Philosophy, religion and the spiritual life/edited by Michael McGhee
  p. cm.—(Royal Institute of Philosophy supplements: 32)
'The present collection of fourteen essays derives from the 1991
Royal Institute of Philosophy conference at Liverpool.'—Introd.
    Includes bibliographical references and index.
        ISBN 0 521 42196 9 (pbk.)
1. Religion—Philosophy—Congresses. 2. Spiritual life—Congresses.
            I. McGhee, Michael. II. Series.
BL51.P547 1992   92-5707
200'.1—dc20   CIP

Origination by Michael Heath Ltd, Reigate, Surrey
Printed in Great Britain by the University Press, Cambridge

Contents

Contents

Introduction

MICHAEL McGHEE

There is a familiar conceptual slope down which philosophers of a certain temper slide when they come to the analysis of religion, and perhaps they do so on purpose, because we like to bring the discussion round to what we can talk about, and if we are sufficiently dominant we become the ones who define what it is proper to say, though the consequence be the stilling of other voices, who may have spoken with understanding.

The philosopher slides from 'religion' to 'religious belief' and from that to 'belief in God', and the latter becomes, imperceptibly, 'belief in the existence of God', so that philosophical reflection about religion is transformed without a pause into reflection on the existence of God, and questions about the rationality of belief, the validity of the proofs, and the coherence of the divine attributes cannot be far behind. It would be absurd to deny the historical importance of natural theology, but there is an established, though slippery, methodology that causes the slide, one that connects natural theology to a certain picture of the procedures that render religious engagement rational or otherwise. The issue, however, is whether such preoccupations should remain central to the philosophy of religion, and, if not, what should replace them.

I

The silent assimilation of questions about 'religion' to questions about 'belief in the existence of God' has already by-passed the Buddhist tradition, which is, in many of its phases, 'non-theistic'. Whether there is anything that corresponds there to 'religious belief' depends on how we construe that term. Certainly it cannot be assimilated to 'belief in the existence of God', or even 'belief in God', though it may in some points be analogous to the latter if we take 'belief in God' as something like Abrahamic faith, the theological virtue, which, in terms of the tradition to which it belongs, is a grounded confidence in the Word. If, for the Buddhist, the impulsion to meditational practice and the re-formation of life arises out of a felt unsatisfactoriness, the first touch of an emergent consciousness, then śraddhā, translatable perhaps as 'faith' or 'confidence', develops pari passu with the confirmation of particular Dharmic claims, internal to the practice, claims which, it is insisted, are

1

Michael McGhee

to be tested, and not merely received or made the object of a 'commitment'. The analogy, if there is one, is with Abrahamic faith, a venture based on a grounded confidence in God's word, and so not assimilable to 'belief in the existence of God'. As Professor Anscombe has made clear (Delaney, 1979, p. 141), 'belief in God' is *believing God*, and the sorts of reason it might be proper to adduce for the latter are hardly the same as those for 'believing in the existence of God', a troublesome expression: philosophers remain divided about this, but there is a tendency now to think that it is a mistake to suppose, as some of our sliding philosophers suppose, that there is something independent of, and prior to, 'belief in God', that the rational agent needs to establish first, viz., 'the existence of God'. Opinion differs thereafter. It may be claimed, for instance, that 'belief in the existence of God' cannot be prised apart from 'belief in God' on the grounds that coming to see that there is a God is *constituted* by recognition and acknowledgment, an immediacy of presence and personal relation, so that there is no separate limbo of bare existential belief. An alternative option is a form of fideism, not of the alleged Wittgensteinian sort. Faith, the theological virtue which waxes or wanes among 'believers', grounds a commitment to an enterprise. Ironically, the commitment aspect of faith has come in some quarters to be directed towards this 'belief in the existence of God', which becomes, though, the object of a *bare* commitment. Thus the assimilation of 'belief in God' to 'belief in the existence of God', the transposition of language appropriate to one context to a radically different one, is apparent in the following passage from Alvin Plantinga:

> . . . the mature theist . . . does not typically accept belief in God tentatively, or hypothetically, . . . (n) or . . . does he accept it as a conclusion from other things that he believes . . . The mature theist *commits* himself to belief in God; this means that he accepts belief in God as basic. (Delaney, 1979, p. 27)

This passage fuses the commitment that manifests faith in the sense of 'believing in God', an attitude of the believer to God, with the supposed attitude of the same believer to the *existence* of God. But whereas the 'belief in God' expressed in a commitment to a venture has religious or spiritual grounds *internal* to a particular tradition, such a commitment, when its object becomes 'belief in the existence of God' *has no grounds*. That is not *necessarily* an objection in itself: but it marks a shift, a difference. In any event, if it *were* a matter of commitment, it would be possible to withdraw from it: but in that case we are not talking about the *foundations* of a person's 'noetic structure', as Plantinga expresses it: the latter is the background against which a commitment is made, not its object. It is one thing, say, to commit oneself to a process of

2

regeneration, acting in the faith that out of death comes life, and quite another to commit oneself to a 'noetic structure', and another still, if it means anything at all, to *commit oneself to believing God*, since believing God is presumably a movement of the heart out of which commitment grows.

This Plantingan fideism is, then, one response to the conception favoured by the philosopher with whom we began. According to this view, the rationality of Abraham's venture of the spirit depends upon establishing a 'real existence': God's. Such a philosopher may readily concede that though it is at least rare to found one's faith on those metaphysical proofs that are, as Pascal said, so remote from our reasonings, it does not follow that the existential propositions implicit in faith cannot be shown to be false, unfounded or incoherent, thus subverting the enterprise of faith for those already engaged.

But it may be a mistake to think in terms of *belief* in the *existence* of God at all, though to say so may lead to charges, not just of fideism, but also of anti-realism and reductionism. How could it be a mistake? One approach might be to say that questions of existence are not so much inappropriate as construed on the wrong model, a model alien to but superimposed upon the real logic of religious discourse. Abrahamic faith, whether construed grossly or subtly, is not dependent upon prior rational deliberation about real existence, nor upon a commitment to a belief-system, and nor is it necessarily undermined by the probing of the culturally received cosmology in which it was originally embedded, though such probing may deeply affect its form. It is a culturally specific form of 'spirituality', formed by as well as shaping the transformations of the local conceptual background which determines the forms which a 'spiritual awakening' may take. Such a claim does not imply that different spiritual traditions are all 'saying the same thing'. Nor does it rule out that such awakenings are to real existence, accessible only to those whose spirit is not sealed by a slumber.

Everything then turns on what may be said to answer to the description 'a spiritual awakening', what such an event may be said to discover, and who is in a position to assess such 'discoveries'. If a new focus of discussion is to emerge in the philosophy of religion, it may be necessary to displace, not just the familiar manoeuvres around 'belief in the existence of God', but the very idea of *belief* as its central concern: (we are not interested in what people *believe*, but in what *insights* are manifested in their lives). This is not a proposal in support of a kind of spiritual non-cognitivism, or a 'religion without doctrines', however, but, on the contrary, a proposal in support of a vision of philosophy as the articulation, the intellectual mapping, of the epistemic inquiry which is an essential strand in the also conative and affective trajectory of the 'spiritual life', a tracking of its transformations and discoveries, in a way which seeks to retrieve the *application* of religious language. To put it another way, displacing belief puts the focus on knowledge, understanding and action. In the Buddhist tradition it is said that one who is 'concentrated' sees things as they really are. Aquinas talks of knowledge of God being a gift of grace belonging only to the good. The point is not that these say the same (they clearly don't) but to inquire into the status of their difference. The suggestion is to establish what the conditions are under which we may come to see what kind of reality such talk attempts to reveal or express, *and what would show such talk to be deluded*. One issue here is whether anyone 'speaks with authority' in religious contexts, whether, through 'spiritual progress', if there is any such thing, a person may come to see the *point* of this or that image, offered by others who have gone before, and so move out of the hearing of those who now stand where they formerly did, but beckoning with the same or similar images. The reference to 'expression' is not 'anti-realist', though it is certainly incompatible with what might contestably be called theological naïve realism. On the other hand, an issue remains, if we stay alert to the greater availability of the Buddhist tradition, about whether we should be seeking to establish a sophisticated realism about God, or whether we should take God-language itself as a local imaging-forth, dependent upon cultural conditions, of what we cannot comprehend. Aquinas said that every way we have of thinking about God is a particular way of failing to understand him as he is in himself. But we might decide that thinking in terms of God at all is a particular form of the breakdown of language in the face of reality, and take refuge in the irony and conceptual modesty of references to dependent origination, an expression which teasingly conceals and unconceals the range and directions of its own applications.

The implication of all this is to see the spiritual life as the ground of concept-formation in religion. In an earlier volume in this series, on the philosophy of religion, Renford Bambrough was critically concerned with 'the main division between those who might be called the grammarians and those who still think of theology and religion as being concerned decisively, though not only, with the world or the universe or reality or how things are' (Brown, 1977, p. 14). But if the grammar had been cartographical, the work of explorers, the distinction would be illusory, though to establish whether it were or no, one would have to follow their tracks. A new form of the philosophy of religion, conceived in these terms, will certainly be exploratory, and, in these dark times, rudimentary. The question will be whether there is anything that the language of our cartographers, those in the stream of the spiritual life, really discloses or unconceals, and, as has been said, what the conditions may be under which their 'disclosures' are to be assessed: 'Phoebus is dead, ephebe. But Phoebus was/a name for something that never could be named.'

So who says so, and who says not?

II

The present collection of fourteen essays derives from the 1991 Royal Institute of Philosophy conference at Liverpool. The Liverpool conference was conceived as an exploratory and interdisciplinary venture, exploratory because it represented an attempt to find and define new ground in the philosophy of religion, and interdisciplinary because it called in aid theologians and other thinkers about religion. It was conceived as an attempt to *start* discussion, as an opportunity for philosophers, theologians and other thinkers to *listen* to a common tradition, and to what each other had to say. Although this is not the place to seek to define differences between theological and philosophical inquiry, it is worth pointing out that the division of labour was by no means always clear cut, in part because many of the participants had some familiarity with what was going on in their neighbour discipline, and also because there were discernible trends on both sides, towards a religious naturalism in some instances, or towards an engagement with the same thinkers, especially perhaps, Plato and Augustine, rejecting or retrieving particular strands of their works. But the Liverpool conference was conceived as exploratory for another reason connected with the very idea of thinking about religion. The participants were invited to consider the relationships between 'Philosophy, Religion and the Spiritual Life', and the invitation extended to reflection on their own experience, if they thought that appropriate. The point was to consider with particular emphasis the role of the 'spiritual life', and give sense to that notion, in relation to religion and in relation to philosophy. This was an attempt to move away decisively from the traditional models of natural theology as the main focus of the philosophy of religion, and to look again at the relations between language, experience and reality, not forgetting to be existing individuals, as Kierkegaard might have put it, and so writing within the discipline of one's own practice and degree of 'appropriation'. The necessity for the 'subjective appropriation' of what might be called 'spiritual truths' is well caught in the British theologian H. A. Williams' remark that 'theological inquiry is basically related to self-awareness and . . . therefore . . . involves a process of self-discovery so that whatever else theology is, it must in some sense be a theology of the self' (1972). This claim does not need to be confined narrowly to theology, of course, but captures something essential to the activity of thinking about religion at all, an activity which anyway draws to itself the scrutiny of the ironist, and rightly so, since a thinker under such a discipline reveals their religious ignorance as well as insight, either in the very stuff of their writing or in the dissonance between their writing and their life, whether or not there is such a thing as religious ignorance or insight. There are other attendant dangers. H. A. Williams' comment needs some qualification if a particular misinterpretation is to be avoided: theological inquiry, or, if we may be more general, reflection on religion, is *related* to self-awareness and to the process of self-discovery, but that is not to say that its subject-matter can be *reduced* to it, and be reinterpreted in terms of human values, human truths, however important they are in their own right. Bambrough had made the point that theology and religion were concerned decisively, though not only, with how things are. We could say that they are also concerned, and equally decisively, with the status and the states of the self, and this for a particular reason connected with the relation between selfhood and reality: to claim that such inquiry is related to self-awareness certainly suggests that such awareness is yet to be established, but there is a further, entirely realist issue, whether the process of self-discovery, suitably understood, is a *condition* of genuinely religious thinking just because it is a condition for the apprehension of how things really are, and not just how they are with the self and its consequent transformations.

III

In the event a surprising diversity and congruence has emerged. There is a wide range of reference, with discussions of the pre-Socratics, Socrates, Plato, Aristotle, Jesus, the Gnostics, Gregory of Nyssa, Augustine, Boethius, Kamalasila, Al-Ghazali, Aquinas, Gregory Palamas and the author of the *Cloud of Unknowing*, Tsong kha pa, Descartes, Kant, Hegel, Kierkegaard, James and Bradley, Heidegger, Lacan and Girard. As far as congruence is concerned, it made for some difficulty in arranging the papers, since there were many points of contact, even between papers apparently unrelated. There were some more or less natural pairs and clusters. For instance, Stephen Clark's re-reading of Descartes in the light of Augustine goes naturally with Sarah Coakley's discussion of Descartes' fourteenth century antecedents in the Christian tradition, but then there are discussions of Augustine in Rowan Williams, Janet Martin Soskice and others. John Haldane's paper on Boethius shares with Anthony O'Hear's an insistence on the importance of aesthetics; Janet Martin Soskice, James Mackey and Michael McGhee all write about ethics.

What follows is not intended to summarize the papers but only to underline the points of contact. Michael Weston's attempt to show that

Kierkegaard remains unsurpassed by later thinkers such as Heidegger seems especially suited to open a collection on philosophy, religion and the spiritual life just because Kierkegaard's work remains for our time a formative critique of the pretensions of the first, the corruptions of the second and the delusions of the third. The papers of John Haldane, Anthony O'Hear and Janet Martin Soskice come together, because not only do they each in different ways bring aesthetics to bear on the issues they discuss, but they all, perhaps because of that, share an insistence on the importance of particulars, over against Platonic universals and noumenal realms. Stephen Clark's paper and Sarah Coakley's are brought together for reasons already mentioned, but Timothy Sprigge's is included with them because he shares with Stephen Clark a decisive rejection of anti-realist tendencies in modern theology. On the other hand, there are points of contact between John Haldane's alertness to the *style* of Boethius and the ancient conception of philosophy as related to the attainment of wisdom or enlightenment (a conception implicit in Timothy Sprigge's Idealist position), and Stephen Clark's retrieval of the Stoics' conception of the role of the philosopher, as contrasted with the modern view of philosophy as a subject, with its unintended though real tendency to corrupt its students. It is worth relating this ancient conception of philosophy to Janet Martin Soskice's situated criticisms of the cultural effects of the notion of an 'unchanging wisdom'. By contrast Sarah Coakley draws attention to the poverty of a philosophy of religion which neglects the *complexity* of the spiritual traditions.

The papers of Ronald Hepburn, James Mackey, Fergus Kerr and Oliver Leaman provide the next sequence, since they could be seen to exemplify a particular line of development. Thus James Mackey responds in effect to Ronald Hepburn's scepticism in the face of recent theologizing about the role of the imagination in religious epistemology by an attempt to discover a meaningful use for the term 'God' through an analysis of the idea of moral obligation conceived, not legalistically, but, interestingly, in terms of a felt impulse or *eros* that we find *within* us but which does not come *from* us. Fergus Kerr's exposition of Girard makes his work on scapegoating available and connects with James Mackey's revision of natural theology since he shows, though not in these terms, how it is precisely the embodiment of this *eros* in the person of Jesus that reveals the futility of the avoidable forces that bring about his death. Whereas Girard shows the possibility of the overcoming of social conflict, Oliver Leaman's account of the relations between ordinary believers, philosophers and mystics in the Islamic tradition describes a case of the actual management of potential conflict within a particular society.

Michael McGhee

Finally, the papers of Paul Williams, Rowan Williams and Michael McGhee complete the sequence, with points of contact around the Buddhist tradition and the nature of self-knowledge. Paul Williams seeks to make available a tradition which is largely unknown among western theologians and philosophers, though fairly generalized references to Buddhism are becoming more frequent. His discussion of the roots of unenlightenment in forms of self-grasping, the idea that particular forms of perception and associated behaviour are the product of fixed and deluded formations of the self connects with Rowan Williams' discussion of the history of self-knowledge and its distortions, in the twentieth century, in early Christianity, in psychotherapy, in 'spirituality', a discussion that has points of contact with most of the other contributions.

References

Brown, S. (ed.). 1977. *Reason and Religion* (Ithaca and London: Cornel University Press).
Delaney, C. (ed.). 1979. *Rationality and Religious Belief* (Notre Dame: University of Notre Dame Press).
Williams, H. A. 1972. *True Resurrection* (London: Mitchell Beazley).

Philosophy and Religion in the Thought of Kierkegaard

MICHAEL WESTON

Kierkegaard is often regarded as a precursor of existential philosophy whose religious concerns may, for philosophical purposes, be safely ignored or, at best, regarded as an unfortunate, if unavoidable, consequence of his complicity with the very metaphysics he did so much to discredit. Kierkegaard himself, however, foresaw this appropriation of his work by philosophy. 'The existing individual who forgets that he is an existing individual will become more and more absent-minded', he wrote, 'and as people sometimes embody the fruits of their leisure moments in books, so we may venture to expect as the fruits of his absent-mindedness the expected existential system—well, perhaps, not all of us, but only those who are as absent-minded as he is' (Kierkegaard, 1968, p. 110). However, it may be rejoined here, this expectation merely shows Kierkegaard's historically unavoidable ignorance of the development of existential philosophy with its opposition to the idea of system and its emphasis upon the very existentiality of the human being. How could a form of thought which, in this way, puts at its centre the very Being of the existing individual, its existentiality, be accused of absent-mindedness? Has it not, rather, recollected that which metaphysics had forgotten? Yet the impression remains that Kierkegaard would not have been persuaded himself that such recollection could constitute remembering that one is an existing individual, for he remarks, of his own ignoring of the difference between Socrates and Plato in his *Philosophical Fragments*, 'By holding Socrates down to the proposition that all knowledge is recollection, he becomes a speculative philosopher instead of an existential thinker, for whom existence is the essential thing. The recollection principle belongs to speculative philosophy, and recollection is immanence, and speculatively and eternally there is no paradox' (Kierkegaard, 1968, p. 184n). We must ask, therefore, whether the recollection of existentiality can cure an existential absent-mindedness or remains itself a form of immanence for which there is no paradox.

I

As this last quotation might suggest, for Kierkegaard, Plato and Hegel represent the beginning and culmination of a particular project of human thought, metaphysics, which, in its claim to reveal the truth of human existence, represents a misunderstanding. This also suggests, of course, that since metaphysics is itself a human enterprise, it thereby misunderstands itself, where the misunderstanding will not be accountable in terms of a failure of metaphysical recollection. But what then is the nature of the latter, and why should it be characterized as 'immanence' for which there is no paradox'? And what indeed is this paradox? 'Our inquiry', says Plato in the *Republic*, 'concerns the greatest of all things, the good life and the bad life' (Plato, 1978, 578c). A man who lived the good life would be *eudaimon*, and *eudaimonia* constitutes the end for our lives: 'We don't need to ask for what end one wishes *eudaimonia*, when one does, for that answer seems final (*telos*)' (Plato, 1975a, 205a). Yet although such an end is, Socrates tells us, 'that which every soul pursues and for its sake does all that it does', we are 'baffled and unable to apprehend its nature adequately' having 'only an intuition (*apomanteuomenos*, announced by a prophet) of it' (Plato, 1978, 505d–e). The human being, unlike the animal, has a conception of his or her own life, that they have a life to live, and so is faced with the question as to what is the good life for themselves. Only in the light of this can they determine the value of different aspects of their lives. But in order to answer this individual question, they must first determine the nature of the good life itself. It is this which humans are unable adequately to apprehend and so remain equally uncertain as to what is the best life for them as individuals. The process of recollection is intended to remove this bafflement, and its nature is revealed in the so-called ascent of the soul in the *Symposium* (201d and following).

This progress is undergone by one who pursues 'beauty in form' and who progressively realizes the nature of the end he is directed towards through the experienced inadequacies of the proposed resolutions. The end proposed by our common bodily nature, physical well-being, is apprehended by the body merely sensuously, both changing with our disposition and lacking any conception of its end in terms of which we could unify our lives. That suggested by our socialized character, social excellence, *arete*, changes as the conventions and traditions of our *polis* or land do and lacks the capacity to say why these *nomoi* should be taken as the ground for the determination of the goal of the individual. The ends proposed by our capacity for knowledge as it reveals itself in the various *epistemai*, forms of knowledge, are multifarious and unable to justify their own primacy in relation to the end sought by human life. Yet such *epistemai* do embody self-conscious procedures of justification and are directed towards the production of truth which ultimately attains a form immune to refutation by contingencies in the timeless truth of arithmetic and geometry. But even here, although the end sought is unchangeable because timeless, such practices are unable to justify either their end *as* truth or *as* the end of human life. That end must both be timeless, and so single and unchanging, and able to bring the questioning as to the end of life to an end. Thus, higher than the particular *epistemai* is the single knowledge concerned with the very nature of knowledge and truth itself (Plato, 1975a, 210e). Such an activity, 'concerned with the final truth, the real nature of things and unchanging reality . . . is the most true knowledge' (Plato, 1975b, 55e).

What, then, is revealed about the desire for the good which finds what it seeks in this activity? Men do not just react to their environment on the promptings of their instinctive desires, but act in the light thrown by a consciousness of their ends. This capacity means that they do not merely live, but have a conception of their lives, and so of a unity through which their lives would express the unity of the 'I', of the soul. It is the *ergon*, the function, of the soul to manage, rule and deliberate (Plato, 1978, 353d), and it can fulfil this function only if it consciously makes of its given nature a unity, becoming 'one man instead of many'. The problem is to identify which of one's capacities is to be given priority so that one's nature as a whole achieves self-conscious unity.

For the human, existence appears as a question, and must therefore be lived in terms of the truth, the answer to the question. But this can only be done if the human knows the truth of the human, and to achieve this he or she must know the truth of truth itself. But then this activity, as human, is itself the resolution of the question, since in engaging in it the human attains to a fully self-conscious ruling of life in terms of truth. 'Those who are uneducated and inexperienced in truth do not have a single aim and purpose in life by which all their actions, public and private, must be directed' (Plato, 1978, 519c). It is because of this that the capacity for rule which constitutes the soul is identical with that of learning and knowledge (Plato, 1978, 581b–c): it fulfils itself in the self-knowledge which is philosophy. The philosopher is, therefore, the consummation of the nature of the human being (Plato, 1978, 490a). Socrates is a man who desires to know whether 'I am a monster more complicated and more furious than Typhon or a gentler and simpler nature to whom a divine and quiet lot is given by nature (*physis*)' (Plato, 1977, 230a). Such knowledge is achieved by knowing the nature of *the problem of human existence* and what can resolve it. That problem is one of self-rule, of making oneself a unity, which can be achieved through giving priority to that capacity whose end is knowledge of the nature of truth itself. Only a life organized in this way is formed knowingly in terms of truth, and so can achieve self-conscious unity: only such a life is truly self-ruling.

If man is to be autonomous, to rule himself, he must know the truth of human life and why it is the truth, and be able through this to unify his own life accordingly. But only the philosopher has, or aspires to, this self-conscious clarity, since his task is precisely to understand the truth of truth itself. Only this activity can constitute the self-conscious ruling of oneself in terms of truth. The notion of the *eidos* itself, it may be said, derives from this, since it is what we can intellectually apprehend in order to rule the world intellectually, to know it in so far as it can be known, and to rule ourselves individually and socially. As what can be intellectually apprehended, it apportions the world and man to the reach of man's capacities. The appearance within Plato's thought of the justification of man's end as contemplation of truth is just that, since its fundamental notions of truth and of the idea and its procedure of recollection, are determined by the desire for autonomy itself. In this sense, the accounts of metaphysics we find in Nietzsche and Heidegger have their validity, since they emphasize that metaphysics results from a particular understanding of human life and so cannot justify it. Recollection here is indeed, as Kierkegaard says, directed towards an immanent solution to the problem of human life.

Now, Plato's answer to the question of the truth of truth is that for there to be the truth about the world and in our thinking both must participate in intelligible form, the former as temporal and spatial instantiations of forms, the latter as recognizing such instantiations as such, and so recognizing, implicitly or explicitly, the forms themselves. It is the relation between the world and the forms which makes the latter '*ousiai*', essences, and so our apprehension of them as of 'truths' rather than as arbitrary definitions, and so this relation, the 'for the sake of which' which unifies the realms of Becoming and timeless Being, the 'Good', is the ultimate principle in terms of which 'truth' can be understood.

For Plato, the truth of our thought about the ideas lies in an adequation between our thinking and these fully intelligible objects, just as the truth of our thought about the world lies in a correspondence between that thought and the things in the world which the ideas make possible. But this relation of correspondence only results in truth because the ideas *are* the truth about the world, that the intelligible form we apprehend is at the same time the form of the world. But what could show that this is so, that our philosophical thought really is true, rather than the drawing out of the presuppositions of a thinking about the world which we non-philosophically take to be so? We could only undermine such a doubt by showing that the notion of truth itself precludes it, so that the question of the possibility of the truth we assume non-philosophically is at the same time the question of the notion of truth itself. The intelligibility of reality as we non-philosophically take it to be for Hegel is not a matter of a harmony, a correspondence, between thought and reality, but a moment in the dynamic which reveals what is true only in the progressive emergence of the notion of truth itself. That notion reveals itself as the identity of subject and object in which reality knows itself, so that its forms as external nature and man's own in his historical development reveal themselves as *for* this absolute knowing. Since absolute knowing has its locus in the human, it constitutes man's own end, his truth: his truth is to be the fulfilled truth of reality as a whole.

The Platonic idea of the Good, the relation of purposiveness which binds the temporal and the intelligible into a whole, is identified by Hegel with the activity of reason itself. The timelessly true is the principle of rationality of the world which comes to its own self-consciousness in human philosophical knowing. It is, therefore, both *ousia* in the Platonic sense, existing 'solely through itself and for its own sake. It is something absolutely self-sufficient, unconditioned, independent, free as well as being the supreme end unto itself' (Hegel, 1984, I, p. 379) and, at the same time, spirit. 'Spirit *is* in the most complete sense. The absolute or highest being belongs to it. But Spirit is . . . only in so far as it is *for* itself, that is, in so far as it posits itself or brings itself forth; for it is only as activity . . . in this activity it is knowing' (ibid. I. p. 143). The rationality of the world is both substance, an intellectually apprehensible order which 'is' in a more than merely temporal sense, and subject, for it is essentially a thinking which must come to know itself. Reality becomes self-transparent in man's absolute knowing. And there man attains true self-consciousness, finding *within himself*, in the form of human history, the ground which can justify his cognitive, practical and political activities, for these represent the concrete manifestations of Spirit's universalizing activity which are for its own self-knowledge. And man can, in absolute knowing, become self-conscious spirit.

Thus, for both Plato and Hegel, man's highest form of activity is philosophical knowing in which the ground, in terms of which all other forms of knowledge and truth can be understood, is discovered as at one with man himself. For Plato, this ground is the idea of the good, of the purposiveness which binds together all that can be said to be and which provides us with the notions of a final truth and unchanging being, and which reveals itself as the ground of the 'divine element' within man, his intelligence, his capacity to participate in the timeless in its *appropriate* form, *as* intelligible, and so reveal the purposiveness which binds the temporal and timeless together. For Hegel, this purposiveness becomes the activity of unifying thought itself, which reveals external nature as *for* the universalizing activity of man's scientific knowing, and man's own as to be imprinted with the image of man *as* such a universalizing being, and so as self-determining. Man, as the being who is capable of knowing his end and acting accordingly, knows his nature as such only in the activity which brings this capacity to fulfilment. And that can take place only in absolute knowing, when external and human nature have been revealed as they *are* through the coming to self-consciousness of the organizing activity of reason. Man does not just possess a divine element, but can in such knowing become God as the ultimate ground of all being, self-conscious Spirit: 'God is spirit, the activity of pure knowing' (ibid. III, p. 283).

II

For metaphysics, the question which our existence is for us is to be answered through a recollection of what is already implicit within that existence, whether this takes the form of Platonic dialectic or the historical dialectic through which Spirit recollects itself. Metaphysics seeks to answer the question by providing a ground, a determination of the nature of the human, through which a concrete form of human life can be justified as life's end as being the fulfilment of that nature. Man, as the being who, unlike the animal, faces existence as a question, must fulfil his nature through a fully self-conscious living in terms of the answer and so in terms of truth. But such self-consciousness can exist only where he *knows* his nature, his Being, and so what justifies it as such: and only philosophical activity within which the truth of truth is known apparently satisfies these demands. Since that activity is recollection, the ultimate ground in terms of which man can understand his own Being and that of all else is implicit within human life itself. Man discovers the ground in terms of which he can live in a truly self-conscious manner within himself and so is capable of autonomy. It is because of this that, referring to Hegel, Kierkegaard remarks in his *Journals* that 'Philosophy is the purely human view of the world, the human stand-point' (Kierkegaard 1970, entry 3253) which tends 'towards a recognition of Christianity's harmony with the universally human consciousness' (ibid. entry 3276). It leads, that is, towards an identification of the human with the divine, a process which has its roots in the Platonic conception of a divine element in man's nature. Hegel's thought, for Kierkegaard, is the culmination of this tradition of philosophy, within which the nature of that human project becomes transparent, for there the human being thinking 'the system of the universe' (Hegel, 1984, I, p. 101) becomes divine. In such thinking he becomes one with self-conscious spirit, and that is God (Hegel, 1984, III, p. 284). Because the problem of existence is one to be resolved by thought revealing what is already implicit there, Kierkegaard says that metaphysics assumes 'that if only the truth is brought to light, its appropriation is a relatively unimportant matter, something which follows as a matter of course' (Kierkegaard, 1968, p. 24).

Kierkegaard's critique of this general project, however, begins with his insistence that it is just this matter of appropriation which *poses* for us the question of the truth of existence: 'the inquiring, speculating and knowing subject . . . raises a question of truth, but he does not raise the question of a subjective truth, the truth of appropriation and assimilation' (ibid. p. 23).

The truth which metaphysics seeks is to be revealed through reflection, and having been apprehended is then to be *lived*. But to say this is immediately to mark a difference between the categories appropriate *within* reflective thought and those concerning our *relation* to it through which it becomes part of our life. 'If a man occupies himself all his life through with logic, he would nevertheless not become logic: he must therefore himself exist in different categories' (ibid. p. 86). These categories are those of 'subjectivity', the relation of the individual to his activities and relationships and so forth which issues from the relation he has to his life as a whole. If an individual occupies himself with logic, we may ask not merely what results ensue but *how* he involves *himself* with it. This question initially prompts an account of the sort of commitment he has to the activity. But this in turn raises a question about that relation itself: is it of the *right* kind? The individual must relate *himself* to this activity as he must to *any* activity or relationship or to anything which occurs to him. Is his form of relationship, then, appropriate for a being subject to such a necessity *throughout his life*? The individual has a conception of his life as a whole, that he has a life to lead, and the question as to the truth of existence relates to this, through which an appropriate relation to activities and relationships *within* life can be determined. But for the individual, his life as a whole cannot be present: 'life constitutes the task. To be finished with life before life has finished with one, is precisely not to have finished the task' (ibid. p. 147). One cannot, therefore, relate to one's life as a whole in terms of a *result* or *fulfilment*, for this is to treat life as a task which can be completed, even if only ideally. But this is precisely what metaphysics does, understanding life's task as the achievement of knowledge of the whole or as the end of the process whereby the whole achieves explicit rationality: 'objective thought translates everything into results, subjective thought puts everything into process and omits results—for as an existing individual he is constantly in process of coming to be' (ibid. p. 68).

Metaphysics in construing life as having an immanent goal fails to recognize that the wholeness of life from the point of view of the *living*, the *existing* individual cannot be so conceived, and it is only from this point of view that the question as to the meaning of life arises. Its view is a result of seeing the question of human life 'objectively', a relation which we as living beings may take up in relation to past human existence, as when we concern ourselves with the objective truth about historical events, but which we cannot take up in relation to our *own*. 'Hegel . . . does not understand history from the point of view of becoming, but with the illusion attached to pastness understands it from the point of view of a finality that excludes all becoming' (ibid. p. 292n). The metaphysical project treats human life in the mode of pastness and only so can it think of it in terms of a final result. But whereas it makes sense to relate to the past in terms of disinterested inquiry and so in terms of the objective truth, such a relation is only possible for a being who has a quite different relation to his own life. 'Whenever a particular existence has been relegated to the past, it is complete, has acquired finality, and is in so far subject to a systematic apprehension . . . but for whom is it so subject? Anyone who is himself an existing individual cannot gain this finality outside existence which corresponds to the eternity into which the past has entered' (ibid. p. 108). His historical inquiry is an activity he engages with and to which he relates: but this latter relation cannot be one of the 'disinterested' inquiry through which he addresses the objects of his research, but one we can only understand in 'subjective' categories. That is, we must understand such a relation in terms of life as it is related to by the one who is *living* it and not in terms of the relation of a living being to a life which is not his own. The comedy of the System, Kierkegaard says, is that it forgets that philosophy has to be written by human beings (ibid. p. 109) who have necessarily a different *kind* of relation to their own lives than they can have to anything else: 'the only reality to which an existing individual may have a relation that is more than cognitive is his own reality' (ibid. p. 280).

How, then, are we to understand existence when it is seen 'subjectively', that is, when it is a matter of an individual regarding his own life? Kierkegaard's answer to this is: as 'becoming'. Whereas objectively life is regarded as if it were in the past, completed and so surveyable by the contemplative gaze of the philosopher, subjectively life is not completable, since one is not done with it until it is done with one. From the existing individual's viewpoint, his own life appears as 'constantly in process of becoming' (ibid. p. 79) and not as directed towards an end. To live, therefore, consistently in terms of this subjective view, 'it is essential that every trace of an objective issue should be eliminated' (ibid. p. 115) and so all trace of living as if an immanently defined end could give significance to one's existence as a whole. To do otherwise is not simply an error of the metaphysical interpretation of life, but characterizes human beings' relations to their lives generally, in ways I shall note in a moment: 'It is enough to bring a sensuous man to despair, for one always feels a need to have something finished and complete' (ibid. p. 79). But to live clear-sightedly in terms of the subjective view, to live as an *existing* individual, is to live one's life *as* constantly in process of becoming, and so not toward an immanent goal (ibid. p. 284). Whereas objectively, one's future is seen in the 'illusion attached to pastness' as if it were directed towards an end surveyable from the present and so *closed*, subjectively the future is open. For the living individual, his future is not already mapped out, tending towards an end: 'The incessant becoming generates the uncertainty of earthly life, where everything is uncertain' (ibid. p. 79). To live related to the essential openness of the future alters too the character of the past. To be so related is to 'strive infinitely' (ibid. p. 84) so that one's concrete activities are not dependent upon the realization of some finite or ideal goal for their significance. As no such goal can have such ultimate significance, the past, whether of achievement or its lack, can have no such significance either, but is merely the base from which one's present striving into the openness of the future takes place. The present, then, is where the past is taken over as one's *own* and so in relation to the absolute openness of one's future. We shall see what this means more concretely for Kierkegaard later.

His critique of metaphysics rests, then, on the contrast between the objective conception of life, where it is seen as if it were already in the past and so complete and surveyable, and the subjective, the way one's life is seen from *within* it, from the point of view of the one who has to live it. And it might appear that Kierkegaard analyses the difference in purely temporal terms. Life as 'becoming' involves, as 'constant striving', the non-ending taking over of one's past into an open future, whereas life objectively conceived is at best progress toward a predetermined future. But it has to be emphasized that Kierkegaard's understanding of these temporal notions is *religious* or ethico-religious: 'all essential knowledge is essentially related to existence. Only ethical and ethico-religious knowledge has an essential relationship to the existence of the knower' (ibid. p. 177). That is, for Kierkegaard, the individual who truly lives as 'becoming' relates to the future as open only in so far as this relation is one to the Infinite or God, and his 'constant striving' constitutes therefore a relation to God, an offering up of his life to the Deity. So that he remarks in the *Journals* (Kierkegaard, 1970, entry 1050): 'To be contemporary with oneself (therefore neither in the future of fear, or of expectation nor in the past) . . . is . . . the God-relationship'. A present within which one takes over one's past in relation to the open future is only possible as such a God-relation since 'the Deity . . . is present as soon as the uncertainty of all things is thought infinitely' (Kierkegaard, 1968, p. 80): that is, the future is only truly open through one's relation to God. And Kierkegaard is far from believing, therefore, that life does not have a *telos*. One who lives his life as always becoming is, because this requires a relation to God, directed towards the end bestowed by God, an 'eternal happiness', although this is, unlike the end understood by the objective views of life, both unattainable through our own efforts and inconceivable as an ideal extrapolation from them and so does not close off the horizon of the future.

I shall return to these notions later. But must not the suspicion immediately arise here that Kierkegaard is involved in re-instating precisely those 'objective' concepts he has shown to be incompatible with the subjective standpoint? Life does not have an end within it, but is now said to have one beyond it. And in that case, life is surely part of an order which is determinate and fixed, even if, unlike the order of metaphysics, it is one we cannot apprehend: 'Reality itself is a system for God; but it cannot be a system for an existing spirit. System and finality correspond to one another, but existence is precisely the opposite of finality' (ibid. p. 107). But given Kierkegaard's critique of the objectivity of metaphysical conceptions, why should the existing individual who understands his existence as constant becoming without a finite end believe in an infinite one, guaranteed by the author of an order beyond our comprehension? Is not this religious construction a last vestige of the hold of objective thinking? Might not we hope to move to a properly existential understanding of existence freed of the metaphysical notions of a determined end within a given order? Certainly Heidegger did.

III

Despite the paucity of references to Kierkegaard in *Being and Time* and Heidegger's earlier works, it is clear that he played a major role in the formation of Heidegger's thought. Heidegger's own account of his relation to Kierkegaard is given in a footnote where he remarks that whereas Kierkegaard 'explicitly seized upon the problem of existence as an existentiell problem, and thought it through in penetrating fashion', he was prevented from an adequate philosophical interpretation of that problem through his adherence to traditional metaphysical conceptions 'the existential problematic' being 'so alien to him' (Heidegger, 1967, p. 494). That is, Kierkegaard thought about the problem of existence as the problem the individual faces in relation to his own existence, and sees certain possible ways in which this may be conceived and resolved by the individual: aesthetically, ethically or religiously. But he is prevented from seeing how the ontic possibilities he discusses are grounded in the Being of human being, and hence from apprehending a more radical interpretation of what those possibilities are and their relation to that Being, through his use of ontological notions which are drawn from intra-worldly beings and are quite inappropriate for the discussion of that being which is in its 'essence', Being-in-the-World.

Heidegger's problem of existentiality concerns the being of human being, that which makes possible the concrete problem or problems we can identify at the individual level, and it is addressed within the context of an attempt to reactivate the question of Being (Heidegger, 1978, p. 41). But that question involves the problem of existentiality only because it is a 'radicalization' of an essential tendency that belongs to man's Being itself (ibid. p. 46). In order for us to relate in any way, theoretically or practically, to 'nature, history, God, space, number', in fact to anything whatever, we must already have an understanding in some way of the Being of these beings (Heidegger, 1982, p. 10). 'Something like Being reveals itself to us in the understanding of Being, an understanding that lies at the root of all of our comportment toward beings' (ibid. p. 16). Such comportment toward beings of whatever kind is a mode of Being of a particular being, ourselves. But whereas other beings have the understanding of their Being in another being, the human, we do not. Rather, we must understand ourselves in our Being: we have an essential relation to our own Being (Heidegger, 1978, p. 54). And as all our comporting towards other beings is something *we* do, it involves at the same time such a relation of ourselves to our own Being. Now, the problem of existence raised at the existentiell level is the question about the meaning of my individual existence, directed towards guiding the conduct of my life. As a relation towards my existence, the question will already involve some understanding of my Being, the Being of human being. The possibility of my conceiving and responding adequately to the existentiell question will depend on the adequacy of my pre-understanding of the Being of the human. And that means, in turn, on the adequacy of my understanding of Being: 'the question of Being, the striving for an understanding of Being, is the basic determinant of existence . . . the question of Being is in itself . . . the question of man' (Heidegger, 1984, p. 16). Of course, it is not necessary for me to be able to *interpret* this understanding of Being: that task is one for philosophy, for the existential interpretation of the Being of man in the service of recalling us to the question of Being. Nevertheless, it remains the case that the existentiell 'problem of existence' depends for its adequate understanding on an adequate pre-ontological understanding of Being. In this way, however radical a departure from traditional philosophical conceptions Heidegger's thought represents, it remains within the context of what Heidegger elsewhere had called the 'human's free appropriation of his whole existence' (Heidegger, 1976, p. 10) which Kierkegaard himself identified as the peculiarly philosophical project: 'the ignorant person merely needs to be reminded in order, by himself, to call to mind what he

19

knows. The truth is not introduced into him but was in him'
(Kierkegaard, 1974a, p. 9). The problem of existence raised at the
existentiell level involves an understanding, although not necessarily
conceptually articulated, of existentiality, the Being of the human, and
that, in turn, of an understanding of Being.

An understanding of our Being is involved in any relating of ourselves to other beings, and so in whatever we do: we always live under
the auspices of some understanding of our Being, of the possibilities of
human being. Such a Being is not something we are related to as
something to be apprehended, but as something to *be*. We are the sort
of being which has its Being to be: we have to live our understanding of
our Being. Since it is through man that the Being of other beings is
disclosed, whilst his Being is disclosed through himself in having to be
that Being, Heidegger calls human being *Da-sein*, *to be* there. As its
Being is always to be for *Dasein*, that Being cannot lie in some determinate state or condition which could be actualized in some particular
Dasein: *Dasein* has always to be its Being and is never finished until it
ceases to exist. And as it always has its Being to *be*, *Dasein* is essentially
concerned about its own Being, which means too that its Being is in
each case *mine*: that each *Dasein* must live itself its own understanding
of its Being.

Nevertheless, although we must always live our understanding of
Being, we do not in the first place and generally do so through an
explicit attention to that Being. Rather our initial understanding lies
implicit in the way we exist prior to any reflective appropriation of
ourselves or of the things we encounter, that way in which we firstly and
for the most part give ourselves immediately and passionately to the
world. *Dasein* is Being-in-the-World. The world is that wherein we
dwell, the familiar environment made up not of things but of relations
of purposiveness. Only in so far as we exist within these relations do we
encounter things at all. In so far as we are in these relations in the
appropriate way, living purposively, the things we encounter, and the
relations themselves, are there in an unobtrusive and unthought
manner.

Since *Dasein* has its Being to be, and so such a Being is always 'mine'
rather than a 'what' which I might merely apprehend, the question
arises as to 'who' this Being is which I make mine. Within the form of
existence which we most immediately are, I take over my Being as a
Being which *anyone* could take over. I am what I do, so that I am my
world, the particular context of purposive relations which is familiar
and mine: I am a shoemaker, teacher, banker, and that others can be
and are. In a similar way I enjoy myself, make judgments, and so on, as
'one' does, in a way available for anyone. I am my Being as something
already given and familiar which can be taken over by me as by anyone:

an already existing environment of modes of work, customs, opinions into which I fit myself. It is only on the basis of such a mode of existing, immediately absorbed in the world of purposive relations, that any more reflective appropriation of my own Being and that of things revealed in the world can take place.

In its immediate form of existing, *Dasein* is simply absorbed in its world, understanding its Being unreflectively in terms of what it can do there. It understands itself, that is, in terms of its success and failure in living within the purposive relations of its world. This inauthentic self-understanding, not drawn from the Being of *Dasein* itself, has an essential temporal structure: *Dasein* awaits the revelation of itself in what the future may bring in terms of success or failure, lives a present absorbed in its world, and has behind it a past which, however much it may be a matter of satisfaction, regret or indifference, is something finished and determinate.

Within everydayness, *Dasein's* self is reflected back to it from what happens or has happened in its world. But this is only possible in so far as *Dasein* is Being-in-the-world, is as absorbed in the purposive relations of world. And it can be so absorbed in these relations, in the in-order-to which reveals equipment and materials, the for-which of the work, and the for-the-sake-of which refers to *Dasein's* potentialities themselves, in so far as *Dasein* itself has an essentially temporal structure, which in the world takes on the particular mode of expecting itself from within its world. But *Dasein* has to *be* its Being, and it has to so long as it is. Hence, its *Being* cannot achieve concretion in any state or condition which could be granted to it by what occurs in the world. Understanding one's Being in this way removes the possibility of understanding it in terms of something to be manifested, in the way of other beings, from within the world. *Dasein* realizes it is not to be identified with any concrete possibility its world offers, but that it is the simple possibility of having to *be* its world. Within the world, and understanding itself not from itself, inauthentically, *Dasein* is at home, in the familiar. Understanding itself from itself, it recognizes itself, however, as '*unheimlich*', not at home and essentially so: as having, not to be *in* its world, absorbed in its familiar relations, but to *be* its world. Such a realization takes place only out of its inauthentic absorption in the world, so that *Dasein* must take over authentically, in terms of its own Being, that Being which it already inauthentically has been.

Such authentic existing, resoluteness, is the pre-eminent form of human temporality. *Dasein* does not await itself in what the future may bring in the world, but always comes toward itself: that is, it takes over what it has been as what must always be taken over, so that the future is open, the past a source of possibility and the present that within which a new revelation of a possibility of its past can occur. *Dasein* takes over what it has been, its world, so that it exists in relation to the world in a new way, in terms of its *own* potentiality. It takes over the concrete possibilities provided by its world but *as* possibilities and not as finished modes of being into which *Dasein* must fit. It appropriates its past not as something finished, but as possibility: and this it can only do by relating to the past as open, by maintaining it as possibility. Since its world is a common one, such engagement with the world constitutes the renewal of a common heritage, a tradition.

To understand one's Being as temporality is to radically distinguish it from the being of intra-worldly beings. It is to realize that one can only *be* that Being as one's Being in the appropriation of the past without issue. And that can only take place if the past is regarded as itself without issue: not as finished and determinate, but as constant possibility of being taken over. To understand oneself in this way is to engage in one's past as an ever renewable source of possibility, to engage in the constant renewal of one's heritage. It is in this that genuinely new creation lies and which enables man to have a history (Heidegger, 1985, p. 279). Man has a history because he is historical: that is, exists as a being which must constantly take over its past as possibility. Of course, for the most part human being must exist inauthentically, simply living at home in the familiar world. But he lives in accordance with his Being, lives it *as* his Being, in creation, in the bringing forth of what is new out of the possibilities made available by his past. In this way he lives his Being as '*unheimlich*', as essentially not at home in the world. Man is not a being among other beings: rather he is as the appropriation of what has been, as *existing* world, within which any other being can have its Being, its own temporal mode.

This may indeed appear as a thinking into the Being of human being which underlies Kierkegaard's criticisms of Hegel in terms of the 'existing individual', a thinking Kierkegaard himself was unable to carry out being still in the grip of certain metaphysical conceptions, but however tempting such an interpretation may be for philosophy, it is not, I think, compatible with Kierkegaard's own thought.

IV

If for Heidegger philosophy has forgotten that Being is a question and so that man is characterized by historicality, for Kierkegaard it has forgotten 'what it means to be a human being. Not, indeed, what it means to be a human being in general, for this is the sort of thing that one might even induce a speculative philosopher to agree to; but what it means that you and I and he are human beings, each one for himself' (Kierkegaard, 1968, p. 109). The philosophical conception of the problem of existence which the individual faces places the conception of the essence of human being at its centre: only if we have an adequate, even if unarticulated, understanding of this can we resolve the problem in the required way. The problem of existence is first and foremost that of 'what it means to be a human being in general', and only through that what it means for me or you to be a human being.

Regarded however from the point of view of the one who has to live it, life appears as a constant task which cannot, therefore, be understood in terms of a fulfilment whether actual or ideal. But if this removes the possibility of thinking of life in terms of an immanent *telos*, it equally removes that of thinking of life in terms of temporal process, although Kierkegaard himself, of course, did not have to consider this philosophical possibility. That possibility remains subject to what Kierkegaard identified as 'the fundamental modern confusion' which 'is to have transformed the communication of capability and oughtness capability into the communication of knowledge. The existential has disappeared' (Kierkegaard, 1970, entry 653). The existential as the relation the individual has to her or his own life must be understood in terms of the *passion* with which life is lived, and in terms of this, the very conception of living in accordance with a notion of the Being of the human is seen to represent an inadequate form of passion, and so one which retains a vestige of the 'objective'. We can see why this should be by briefly reviewing the three main forms which human existing can take according to Kierkegaard, the aesthetic, the ethical and the religious.

In relation to the question which one's life is for oneself, one may look towards the achievement and maintenance of some condition constituting an overriding goal for the rest of one's life, as power, position or wealth, or, more modestly, towards some set of material and personal circumstances with which one could be content. Either way, however, these conditions have their significance for us only in so far as we desire them. The aesthetic view of life tries, that is, to understand what gives meaning to our lives as a whole as lying in the dispositions with which we find ourselves. Nevertheless, as facts about ourselves, these are subject to change: if we cease to desire riches, say, then we will alter our understanding of where life's meaning lies. What gives significance to the particular proclivities that at any time fill this role, therefore, does not lie within them, but is rather given to them by ourselves. Yet, at the same time, such a granting of significance within the aesthetic condition is merely a response to the dominance of some inclination we find ourselves with, so that as this changes so does our conception of what is important in life. Within the aesthetic form of existence, that is, I must determine what proclivity is to play this role, whilst, at the same time I *evade* recognition of this necessity: I find the significance of my life *determined* in a certain way, yet decline to recognize the essential role of my consent in this. Aesthetic existence involves, therefore, a self-deception: I have always chosen, but the very mode of my life, lived in terms of the satisfaction of my given dispositions, seeks to deny this. If I reflect, however, I realize both that whatever proclivity has played the dominant role in my life has not been explicitly chosen in terms of its adequacy to resolve the question, of the meaning of life, which is to be resolved, and that *no* given disposition *could* answer this, since the facts as to what is given may change at any time but the question would remain.

Since it is my capacity for choice which gave their role to my given dispositions in determining life's significance, it may appear that the question can only be resolved if I choose, not in terms of their predominance at any time, but in terms of choice itself. The ethical individual 'can impart to (his history) continuity, for this it acquires only when it is not the sum of all that has happened to me but is my own work' (Kierkegaard, 1959, II, p. 255). He or she has choice itself as the measure for life as a whole. I choose, that is, not in terms of what my dispositions at any time suggest and so as directed towards the achievement and maintenance of certain conditions, but in terms of my capacity for choice itself, and so unconditionally. The resolution to the problem lies in impressing whatever is given, both by one's nature and by what happens in life, with oneself: 'The great thing is not to be this or that but to be oneself' (ibid. II, p. 180). This is done by choosing oneself, and so rendering oneself independent of external or internal conditions which may or may not come about. One undertakes unconditional choices which raise aspects of one's given nature to the level of things chosen by oneself: marriage in relation to one's sexuality, vocation in relation to one's talents, and so on. And one freely appropriates everything which happens in one's life, both the joyful and the sorrowful. In this way, by imposing unconditional choice upon one's life, 'we win through to an entirely individual human being' (Kierkegaard, 1968, p. 227), since in so doing I impress the facticities of my life with the very *form* of the 'I' which I am throughout my life. Whereas the categories of the aesthetic view are those of fortune, misfortune and fate, those of the ethical are victory and defeat, of myself as self-determining over my factually given dispositions and the events of my life.

But can this truly be the answer to the question of what can give significance to my life *as a whole*? Is there not rather a hidden complicity between the aesthetic and the ethical views of life? The very structure of the ethical project, to gain a victory over myself, indicates that there is. The ethical determines what can give meaning to my life as a whole in terms of my capacity for choice, for imposing unconditional choice upon myself and holding to it. But this, in so far as it is a capacity which I have, is still a *part* of my life, a part of that which is to be given meaning as a whole, just as is any dominant desire within the aesthetic view. The ethical still conceives life objectively, in terms of a goal whether factual or ideal, projected by man's own capacities. But no such part can give meaning to the whole, for what is it which gives meaning to my exercise of freedom?

The question which human life faces us with is, as far as life itself is concerned, paradoxical: life cannot determine its own significance in terms of itself. This realization compels the recognition that meaning can only be given to one's life as a whole by relating it *as* a whole to an *absolute* Good. An absolute Good is one to which I can relate my *whole* existence, in terms of which I recognize that nothing I can do, and so no capacity I may or could possess, and nothing that happens to me can give meaning to my life as a whole. One can know nothing concrete about such an absolute Good (Kierkegaard, 1974a, p. 55) since it does not lie in the exercise, fulfilment or result of any human capacity, except that to relate to it requires that one wills one's life in its totality absolutely, unconditionally, without looking for any result, and hence for anything, including one's victory over oneself, which one would deem good. And this means that one's activity in relation to this good *cannot* be regarded as means towards its achievement. Rather, recognition of this absolute good can, in so far as it results in one's activity, only take the form of the *negative* movement of removing within one's life the illusion of a humanly projected goodness, and so ultimately of total self-renunciation: 'in self-renunciation one understands one is capable of nothing' (Kierkegaard, 1962, p. 355).

This 'absolute Good' for the individual is what Kierkegaard means by an 'eternal happiness'. But this notion cannot play a similar role to that of 'happiness' within the aesthetic and ethical forms of existence. There it would make sense to ask in what happiness consists, for it would be used to refer to some state either attainable *within* life or as an ideal which one must conceive in order to pursue. But the notion of an 'eternal happiness' merely identifies 'the good which is attainable by venturing everything' (Kierkegaard, 1968, p. 382) in relation to which our activity cannot be the utilization of means towards the achievement of an end. Therefore, Kierkegaard says 'the resolved individual does not even will to know anything more about this *telos* than that it exists, for as soon as he acquires some knowledge about it, he already begins to be retarded in his striving' (Kierkegaard, 1968, p. 353). To will to 'know' something about it is to construe the 'absolute Good' as if it were something we could achieve or approach of our own powers and so need a prior conception in order to direct our activity. But the absolute Good is that which requires us to venture everything and only so can it give significance to one's *whole* life. It cannot therefore be construed as within the reach of our powers, whatever they may be, which are necessarily a *part* of the life which is to be given up in its *entirety* to this good. What could give such significance must, therefore, have for us an essentially *negative* form: all we can know about it is that it cannot be known and all we can say about it is that it requires us to venture everything.

An 'eternal happiness' marks this absolute *telos* for the individual, his absolute Good. The Goodness of this good is that to which the individual must relate in order to have his own absolute end, and this is the notion of absolute goodness itself, God. 'God is a highest conception, not to be explained in terms of other things, but explicable by exploring more and more profoundly the conception itself' (Kierkegaard, 1968, p. 197). That absolute Goodness, the measure for my life *as a whole*, can, that is, be related to only in 'the mode of absolute devotion', and that to which we can be so related is 'God'. 'Self-annihilation is the essential form for the God-relationship' (ibid. p. 412). I cannot, therefore, relate to God through my understanding, as if I could *grasp* the measure for my life as a whole and its reason and so set about making my life in accordance with it, since then my activity would be a means towards the achievement of the individual good determined by that measure. God is, seen from the point of view of reason, the 'limit to which reason repeatedly comes', the 'unknown with which reason collides' (Kierkegaard, 1974, I, p. 55) since, as the measure for life in its entirety, all reason can know is that God requires, *as* God, the *whole* of life, and therefore the recognition by the individual 'that he is nothing before God, or to be wholly nothing and to exist thus before God' (Kierkegaard, 1968, p. 412). The relation to God requires the sacrifice of our reason and understanding in the sense of a giving up of their claim to be able to establish and reveal a measure for life as a whole which human powers may achieve or advance towards: 'The contradiction which arrests (the understanding) is that a man is required to make the greatest possible sacrifice, to dedicate his whole life as a sacrifice— and wherefore? There is indeed no wherefore' (Kierkegaard, 1972, p. 121). The limit of our reason is to reveal this contradiction, the impossibility of a human resolution to the problem of the meaning of existence, and that the relation to the absolute measure requires the active giving up of such presumption. Such an understanding of existence, which really does encompass life as whole, is, therefore, according to Kierkegaard, essentially religious.

The apprehension of the relation to God as constituting *the* meaning for human life is the religious understanding. It has two forms for Kierkegaard. The universally religious, or 'religion A', understands what this requires in human terms, a turning away from all humanly determined goods as constituting *the* good. It thus conceives of the demand as one of 'infinite resignation', an active offering up of human life as it is lived to God. This involves seeing the task of life as one of the constant exercise of the renunciation of absolute concern with finite results, and so of dying away from the world: 'the individual who sustains an absolute relationship to the absolute *telos* may very well exist in relative ends, precisely in order to exercise the absolute relationship in renunciation' (Kierkegaard, 1968, p. 367). Christianity, or 'religion B', goes beyond this, by involving Faith, the belief that, having offered one's self to God in the striving of infinite resignation, it will be given back, so that one's life is no longer characterized by striving against one's tendencies to will relatively, but by an absolute purity within the world. 'Faith, after having made the movements of infinity . . . makes those of finiteness' (Kierkegaard, 1974b, p. 48). Christianity is the 'absolute religion' for Kierkegaard because it recognizes in its most radical form the difference between man and God (Kierkegaard, 1970, entry 46), the nothingness of man before God: that nothing that man does, not even 'infinite resignation', can have any value unless it is given by God to man. And the transformation of the individual to absolute purity is not something which the individual can accomplish of himself, since it involves a *total* transformation of the self away from relative to absolute willing.

We can now see why Kierkegaard stresses the centrality, in relation to the problem of existence, of the *fact* that *I* exist, rather than of a conception of the nature of human being. The problem of existence is faced by the individual in having to live his life *as* a whole, as *the* life which *he* or *she* has to live. This problem cannot be addressed by reference to some aspect of human existence in terms of which the individual could organize his life, since any such aspect is only a part of what must be given meaning as a whole. Of course, philosophy attempts to deal with this by trying to show how such a mode of existence fulfils human nature, 'what it means to be a human being in general'. But the pursuit of an essence of human nature, even if this opposes a metaphysical conception of Being, is motivated by the understanding that the problem of existence at the individual level *can* be resolved by referring to some particular mode of existence as an exercise of human powers, since to speak of the Being of human being is precisely to speak of what could justify appeal to such a mode in solving that problem. But if reference to *no* such aspect of an individual's existence, neither, for example, his intelligence, his reason, nor his capacity for creation, can *in principle* resolve that problem, then the pursuit of the Being of human being is misplaced. The religious understanding claims that the problem is such that this is indeed the case, that the individual can only relate to his life *as a whole* in recognizing an absolute Good to which life *in toto* can be given: and one can only express such a recognition in self-renunciation. Human being does not have an essence, a Being, for his existence is a problem which *cannot* be resolved by referring to such a Being. It is this philosophy forgets: 'what it means that you and I and he are human beings, each one for himself' (Kierkegaard, 1968, p. 109). It forgets the relation the individual has to his own life, and so the subjective categories concerned with the passion with which life is lived. The problem of human life is of how the individual can *commit* his life in its entirety, and the resolution of this precludes the very *form* of the philosophical response which attempts, in terms of a determination of man's Being, to give a privilege to certain of man's own powers. Kierkegaard does not give an account of a problem at the existentiell level which can only properly be addressed through an adequate understanding of the existentiality, the Being of human being, since no such understanding could allow a resolution of the problem that account is directed towards. What that problem requires is giving up the *presumption* of such an understanding to resolve it, since it requires, quite simply, the giving up of *all* human presumption to be able to give meaning to human life. It is not merely that religious existence is the 'mortal enemy' of 'a human's free appropriation of his whole existence' (Heidegger, 1976, p. 21) but rather that the problem which human existence is shows that no such 'free appropriation' is possible.

References

Hegel, G. W. F. 1984. *Lectures on the Philosophy of Religion*, Vols. I–III, P. Hodgson (ed.). (University of California Press).

Heidegger, M. 1967. *Being and Time*, trans. J. MacQuarrie. (Blackwell).

Heidegger, M. 1976. *The Piety of Thinking*, trans. J. G. Hart and J. C. Maraldo. (Indiana University Press).

Heidegger, M. 1978. *Basic Writings*, ed. D. Krell. (Routledge).

Heidegger, M. 1982. *The Basic Problems of Phenomenology*, trans. A. Hofstadter. (Indiana University Press).

Heidegger, M. 1984. *The Metaphysical Foundations of Logic*, trans. M. Heim. (Indiana University Press).

Heidegger, M. 1985. *The History of the Concept of Time*, trans. T. Kiesel. (Indiana University Press).

Kierkegaard, S. 1959. *Either/Or*, Vols. I and II, trans. W. Lowrie. (Anchor).

Kierkegaard, S. 1962. *Works of Love*, trans. H. and E. Hong. (Harper).

Kierkegaard, S. 1968. *Concluding Unscientific Postscript*, trans. D. Swenson. (Princeton University Press).

Kierkegaard, S. 1970. *Journals and Papers*, ed. and trans. H. and E. Hong. (Indiana University Press).

Kierkegaard, S. 1972. *Training in Christianity*, trans. W. Lowrie. (Princeton University Press).

Kierkegaard, S., 1974a. *Philosophical Fragments*, trans. D. Swenson. (Princeton University Press).

Kierkegaard, S. 1974b. *Fear and Trembling*, trans. W. Lowrie. (Princeton University Press).

Plato, 1975a. *Symposium*, trans. W. Lamb. (Heinemann).

Plato. 1975b. *Philebus*, trans. W. Lamb. (Heinemann).

Plato. 1977. *Phaedrus*, trans. H. Fowler. (Heinemann).

Plato. 1978. *Republic*, trans. P. Shorey, (Heinemann).

De Consolatione Philosophiae

JOHN HALDANE

I

While I was quietly thinking these thoughts [about misfortune] over to myself and giving vent to my sorrow with the help of my pen, I became aware of a woman standing over me. She was of awe-inspiring appearance, her eyes burning and keen beyond the usual power of men. She was so full of years that I could hardly think of her as of my own generation, and yet she possessed a vivid colour and undiminished vigour . . . Her clothes were made of imperishable material, of the finest thread woven with the most delicate skill . . . On the bottom hem could be read the embroidered Greek letter *Pi*, and on the top hem the Greek letter *Theta*. Between the two a ladder of steps rose from the lower to the higher letter. Her dress had been torn by the hands of marauders who had each carried off such pieces as he could get. There were some books in her right hand and in her left hand she held a sceptre . . . As she spoke she gathered her dress into a fold and wiped from my eyes the tears that filled them . . . the clouds of my grief dissolved and I drank in the light.[1]

So begins the famous work by Boethius from which I have taken my title: a work which, according to one of its earliest translators, King Alfred, is 'among the books most necessary for all men to know'. The woman of Boethius' vision is, of course, *Philosophy*—'the nurse in whose house I had been cared for since my youth'.

That this general style of writing is not one favoured by present-day academic philosophers may seem to be a relatively insignificant fact connected with changing literary traditions. After all, in the course of the work Boethius presents or presupposes views on such perennial matters as the mind/body problem, knowledge, free-will, general ontology and time. I am sure, however, that the marked stylistic contrast also betokens a difference of belief about the nature and role of philosophy. Boethius' literary form is shaped by, and gives further shape to, an ancient conception of philosophy according to which its *telos* is the attainment of wisdom, philosophical enquiry being the discernment of, and progression along, a pathway leading to enlightenment, and *ipso*

[1] Boethius (1987), I, prose 1 & 3, pp. 35–9.

facto to what St Augustine termed *gaudium de veritate*—delight in the attainment of truth.

Viewing things in this way it also becomes clear why there is a problem in marking a sharp distinction within the writings of antiquity and the medieval period between philosophical texts and works of spiritual counsel. That we have no difficulty applying such a distinction to contemporary writings reveals much about the shared self-conception of present-day professional philosophers. What I have to say here may reflect some light upon the source of this difference. However, my ultimate concern is with a philosophical issue rather than with historical or interpretative ones, though those will feature along the way. More precisely, I want to raise two questions: first, can any interesting sense be made of the claim that philosophy might be a source of comfort or consolation? and second, is there reason to believe this claim to be true?

Let me add straight away that, so far as I am concerned, these questions introduce a genuinely open inquiry. I do not have any settled conviction on the matter, but equally, I should not be troubling to consider these issues if I did not think that the questions might be answered positively, and believe that it is important that we should know whether this is indeed so. There is, I believe, no single correct answer to the prior question *what is philosophy*? but if one of the things it *cannot* be is a source of reasoned comfort in circumstances where experience and reason itself lead us to be troubled about the human condition, its contingency and its evils, then this is something that needs to be demonstrated, appreciated and proclaimed—perhaps in the style of Quine who writes of even aesthetics and moral philosophy as 'being apt to offer little in the way of inspiration or consolation' (1981, p. 193).

II

Although I am not principally concerned with exegesis of Boethius' text, I want to consider certain ideas to be found there and in other of his writings, and to relate these to questions concerning the interpretation of a broad class of thoughtful experiences of nature and of art. It is a matter for some doubt, however, whether Boethius' ideas constitute a coherent whole. One may reasonably question whether what he has to say in *De Consolatione* is compatible with the sort of metaphysics he explores in *De Trinitate* and in his second commentary on Porphyry's *Isagoge* (an introduction to Aristotle's *Categories*). What I have in mind in saying this is that in the first and third of these works he offers a version of neo-Platonism in which the forms of things subsist in the mind of God, independently of and prior to their empirical instances; whereas in his commentary on Porphyry he sets up the problem of universals as being that of how natures can be both many and one, and then rehearses (and commends—though perhaps without endorsing) a recognizably Aristotelian solution: members of a kind have numerically distinct sensible natures but when they are brought under conceptual investigation a common universal nature is discerned by the intellect. I mention this instance of philosophical tension not just by way of example but because I shall have particular reason to return to it later. For now, however, I need to say more about the central theme of Boethius' essay and his treatment of it.

De Consolatione is a complex work almost every aspect of which presents interpretative questions. It was written in the period prior to his execution for treason, and the historical background, at least, is fairly straightforward. Boethius was born sometime after 480 into an aristocratic Christian Roman family. His father, who was a prefect of Rome and a consul, died when Boethius was a boy and this led to his being adopted by Quintus Symmachus. His new family was even more eminent than his native one, and thereby Boethius was confirmed as a member of the highest Patrician class in a period when the loyalties and virtues of that group were tested by service to Gothic rulers in Rome and by divisions within the western and eastern branches of the Church. He was evidently a juvenile prodigy, exhibiting talents for scholarship and administration. The latter skill drew him to the attention of King Theodoric the Ostrogoth who quickly promoted him a consul and thereafter gave him the offices of *magister officiorum* and *magister dignitatis*—in effect, head of the civil service and of the royal court.

During his youth and the period of his early political career Boethius studied the available works of Plato, Aristotle, Cicero, Porphyry and Augustine (hence his remark about having been 'cared for in the house of philosophy'). Perhaps in recognition of impending cultural disaster he set himself the task of translating and writing commentaries on the major Greek philosophical texts; but a further, and presumably more important, motive for this work was his belief that the central ideas of Plato and Aristotle are mutually compatible sets of truths which provide the philosophical means for articulating Christian doctrine.

Following political and religious difficulties between the western and eastern branches of the Empire, Boethius was accused of treason, exiled and imprisoned in Pavia (in northern Italy) where he was brutally tortured and beaten to death in 524. It was during the period of his exile that he composed *De Consolatione*. This is largely in dialogue form, being divided into five books each of which is sub-divided into sections containing prose and poetry. The work begins with Boethius bemoaning this wretched fate in muse-inspired verse:

I who once composed with eager zest
Am driven by grief to shelter in sad songs;
All torn the Muses' cheeks who spell the words
For elegies that wet my face with tears
No terror could discourage them at least
From coming with me on my way.
They were the glory of my happy youth
And still they comfort me in hapless age.[2]

As he reflects on what he has written, however, he becomes aware of the figure of a woman—the Lady *Philosophy*—and there then follows the description of her with which I began. Like the entire work, this description is rich in allusion: the Lady is aged but of 'undiminished vigour' (perennial wisdom), wearing a self-woven garment of imperishable cloth (incorruptible and autonomous) with an embroidered ladder leading from *Pi* to *Theta* (the continuous ascent from practical to theoretical philosophy), but showing signs of abuse (schism within philosophy itself).

Philosophy's first words concern the Muses from whom Boethius had been taking comfort:

Who are these hysterical sluts to approach this sick man's bedside? . . . These are the very women who kill the rich and fruitful harvest of Reason with the barren thorns of Passion. They habituate men to their sickness of mind instead of curing them . . . be gone, and leave him for my own Muses to heal and cure.[3]

These harsh words recall something of Socrates' criticisms of Homer,[4] and they are shortly followed by a direct reference to the dialogues in which Boethius' innocent predicament is likened to that of Socrates and, perhaps (through the use of the phrase 'a victorious death'), to that of Christ also.[5] So the healing cure begins. Following his rehearsal of his troubles and misery, Boethius is urged to recall that his nature transcends that of a rational mortal animal, and likewise that the ultimate reality is not the empirical order but the beginning and end

[2] *Consolation*, I, poem 1, p. 35.

[3] *Consolation*, I, prose 1, p. 36.

[4] *Republic* 595 b9–c3 and 607 c3–8. Also St Augustine, see *De Doctrina Christiana* ii, 18, 28.

[5] *Consolation*, I, prose 3, p. 39. The parallel with Christ is suggested in passing by Henry Chadwick in his excellent book *Boethius: The Consolations of Music, Logic, Theology, and Philosophy* (Oxford: Clarendon, 1981).

(i.e. goal) of its existence, *viz.*, God. Thus misery at worldly misfortune involves a double error: identifying the self with the embodying organism, and reality with the sensible world. Whereas the states of mind induced by the arts encourage this misidentification, by directing attention towards sensuous forms, the effect of philosophy is to encourage composure and detachment by reminding the soul of its true nature and origins.

III

There are scarcely any present-day philosophical discussions of the stated theme of *De Consolatione*.[6] But recently an article was published by Andrew Belsey in which he discusses various ironic aspects of the work and addresses the question of what view of the philosopher's vocation emerges from it (1991, pp. 1–15). Belsey brings certain broadly political interests to his reading of Boethius and worries at the apparent encouragement of retreat into philosophical contemplation. Throughout the work, however, he discerns various levels of irony in the public recommendation of private detachment, and so finds it possible at least to qualify the charge of self-indulgent stoicism in the face of evil:

> What was said in the act of saying, even if not in the saying itself, was true philosophy—the transcending of disengagement, the metamorphosing of private consolation into political action. Far from being a turncoat, Boethius would be a beacon illuminating the true philosophical path (p. 14).

Belsey is properly alert to the likelihood of literary subtleties and paradoxes in the text, but I think there are philosophical complexities from which his attention is distracted by his moral-cum-political interest. As regards the question of whether Boethius is offering a basically contemplative solution to what I shall just call 'the problem of life', I think the answer is 'yes', but not in the form in which contemplation is first introduced (as a form of Stoical detachment), which is where Belsey focuses a good deal of his interest. Moreover, I think that if anything can be made of the idea of a philosophical solution to 'the problem', it will require a metaphysical outlook that is at once *more substantial* than the gaze of 'calm indifference' which Belsey finally allows—so long as it is accompanied by combat with 'the gross and contemptible evils of worldly power' (p. 15)—but also *less substantial* than that endorsed by Boethius himself.

[6] I mean by this to exclude the literature on Boethius' discussions of divine foreknowledge and human freedom.

The dialogue is presented as the record of a course of treatment in which Boethius is given a series of increasingly more powerful cures. At each stage the condition of his soul is revealed by the character of his complaints against the order of things, and a cure is administered relative to that condition. There are, then, several forms of philosophical consolation, and it is important to see that while these overlap (to some extent), they also draw upon two distinct traditions, *Stoicism* and *neo-Platonism*, and differ in their metaphysical commitments and practical implications.

In *Book II*, Boethius is reminded that he has been as much the beneficiary of Fortune as her victim, and therefore has no grounds for complaint that she has dealt him an unwarranted blow. Nothing is deserved and no objection or resentment is in order when life takes an unwelcome turn. Knowing this, one should bear one's gains and losses with equanimity and hold fast to the only enduring goods: self-possession of one's soul, the knowledge that the course of events is under the direction of a cosmic force, and the love of true friends:

> The world in constant change
> Maintains a harmony,
> And elements keep peace
> Whose nature is to clash . . .
> If Love relaxed the reins
> All things that now keep peace
> Would wage continual war
> The fabric to destroy
> Which unity has formed
> With motions beautiful.
> Love, too, holds peoples joined
> By sacred bond of treaty . . .
> O happy race of men
> If Love who rules the sky
> Could rule your hearts as well![7]

Although in his survey of the text Belsey notes other elements of consolation, the attitude he focuses upon is that induced by this phase of the treatment, namely Stoical 'enlightened acceptance'. But immediately following the verses quoted above, *Philosophy* passes on to the next stage of treatment, saying 'the remedies still to come are, in fact, of such a kind that they taste bitter to the tongue, but grow sweet once they are absorbed'.[8] This introduces a discussion of the varieties of happiness and a specification of the 'true form' which the soul has the

[7] *Consolation*, II, poem 8, p. 77.
[8] *Consolation*, III, prose 1, p. 78.

power to attain by means of philosophy—again Augustine's *gaudium de veritate*. Two ideas appear early on in this *Book* which signal the development of a neo-Platonic and non-Stoic form of consolation. These are the doctrine of knowledge as recollection (*anamnesis*) and the notion that earthly unhappiness is a result of the soul's embodiment—like a captured song bird reduced to whispered airs of longing for the woodland home.

From that point onwards the treatment accelerates with ever-more potent cures. It is argued that certain qualities are pre-eminent values, and that they are extensionally equivalent: 'sufficiency, power, glory, reverence and happiness differ in name but not in substance'.[9] There is then a pre-figuring of Anselm's ontological proofs for the existence of a maximally great being, i.e. one possessed of these 'different but identical' qualities in their highest degree of perfection. Since what all things desire is their happiness, and perfect happiness is to be found in God, the task of philosophy becomes that of launching the soul into flight towards the still point of the turning worlds. There, at the invisible centre, is a divinity from out of whose intellect emerged all things corporeal and incorporeal. Since the soul is (or, more strictly, contains) intellect it has a co-natural affinity with the source of its being, and to the extent that it comes to be possessed of the same perfections as constitute both the divine essence and what that essence (in its guise as *Intellect*) contemplates, i.e. ideal forms, it is to that degree itself divine: 'Each happy individual is therefore divine. While only God is so by nature, as many as you like may become so by participation'.[10]

Philosophy has more to say in reply to Boethius' questions about fate and the existence of evil in a world ordained by Providence. Basically, the story is that Fate is the working out of the divine order through the causal nexus, and it is an aspect of divine activity which enlightened intellects stand on the far side of, viewing it with satisfaction and not, as do the ignorant, with resentment. The effects of Providence are only Fate for those who do not or cannot understand them. Likewise, the appearance of evil is an illusion, everything that happens has point and works ultimately for the greater glory of *Nous*, or, as Boethius might more likely have said, *ad maiorem dei gloriam:*

> If you desire to see and understand
> In purity of mind the laws of God,
> Your sight must on the highest point of heaven rest
> Where through the lawful covenant of things
> The wandering stars preserve their ancient peace: . . .
> Those things which stable order now protects,

[9] *Consolation*, III, prose 9, p. 95.
[10] *Consolation*, III, prose 10, p. 102.

Divorced from their true source would fall apart.
This is the love of which all things partake,
The end of good their chosen goal and close:
No other way can they expect to last,
Unless with love for love repaid they turn
And seek again the cause that gave them birth.[11]

What has now been reached is far from a stoical philosophy of passive reconciliation. While it suggests that the philosophical life will be an active one, Belsey is right to imply that what Boethius commends is not his own preferred form of engagement, i.e. combatting 'the gross and contemptible evils of worldly power'. But nor, I think, is it quite that which Belsey actually attributes to Boethius in writing *De Consolatione*, i.e. a combination of metaphysics with 'the transcending of disengagement, the metamorphosing of private consolation into political action . . . [and] the continuation of engagement even to the tragic end'. For Boethius, the essence of *philosophy* consists in the perfecting of one's intellect, thereby participating in the eternal contemplative life of God. That is its consolation: the perfect quality of its object, and *ipso facto* of itself and of its accomplished practitioner. As Socrates has it, reason is best because its objects are most noble.[12] What this account involves are several decidedly Platonistic elements and processes. Some of these have already been indicated but another which it is relevant to mention is the idea that our general concepts of things, such as those of cats, bats and rats are not only innate but *could not* have been acquired from empirical experience of instances of these natures, since, strictly speaking, there are no such instances but only 'images' of natures:

> [F]rom those forms which are outside matter come the forms which are in matter and produce bodies. We misname the entities that reside in bodies when we call them forms; they are mere images; they only resemble those forms which are not incorporate in matter.[13]

IV

What, then, is to be made of Boethius as a source of consolation? Compounding my earlier questions, has he an intelligible and plausible view? If this means to ask whether someone who was troubled by 'the problem of life' might reasonably take comfort from words spoken by

[11] *Consolation*, IV, poem 6, pp. 141–2.
[12] *Republic*, 509 d6–511 e5.
[13] *De Trinitate*, II, as translated by H. Stewart and E. Rand, *The Theological Tracts*, Loeb Classical Library, *Boethius* (London: Heinemann, 1926).

the *Lady Philosophy*, then I think the answer is 'yes'. The work contains much enduringly good spiritual guidance of a mundane sort about recognizing contingencies, not exaggerating one's misfortune, counting one's blessings and discounting trivial goods. Also, notwithstanding the Platonic demotion of the sensible world to the domain of images and the banishment of the Muses, there are some striking poetic reminders of natural beauty which prefigure Hopkins in both their style and spirituality:

The flower-bearing year will breathe sweet scent,
In summer torrid days will dry the corn,
Ripe autumn will return with fruit endowed,
And falling rains will moisten wintry days.
This mixture brings to birth and nourishes
All things submerged in death's finality.
Meanwhile there sits on high the Lord of things,
Who rules and guides the reins of all that's made,
Their king and lord, their fount and origin,
Their law and judge of what is right and due.[14]

None of this, however, amounts to a *consolation of philosophy*—as opposed to wise and welcome words from a philosopher. If one looks then to the Boethian philosophy itself, I think it may be intelligible that one should regard the possibility of participation in the life of a Divine intellect as not merely comforting but as an all-consuming purpose of existence which would exclude the very possibility of knowledgeable misery. But while it may be coherent, it barely registers on the credibility measure. It is, at best, an intelligible but implausible hypothesis. Moreover, it is even unclear what would count as phenomenological evidence in its favour. What, if anything, would it be like to intuit the essence of catness? or more generally to comprehend being as such? And if the answer is that there is nothing it is experientially 'like' for

---

[14] *Consolation*, IV, poem 6, p. 142. Compare this with, for example, Hopkins' poem 'God's Grandeur' (in particular the second stanza): The world is charged with the grandeur of God./It will flame out, like shining from shook foil;/It gathers to a greatness, like the ooze of oil/Crushed. Why do men then now not reck his rod?/Generations have trod, have trod, have trod;/ And all is seared with trade; bleared, smeared with toil;/And wears man's smudge and shares man's smell: the soil/Is bare now, nor can foot feel, being shod./And for all this, nature is never spent;/There lives the dearest freshness deep down things;/and though the last lights off the black West went/Oh morning, at the brown brink eastward, springs—/Because the Holy Ghost over the bent/World broods with warm breast and ah! bright wings. *The Poems of Gerard Manley Hopkins*, ed. W. H. Gardner and N. H. MacKenzie 4th edition (Oxford: Clarendon, 1970), p. 66.

one's intellect to grasp ideas in the mind of God, then I think the problem arises of how, even if this were a possibility, *we*—as we find ourselves in the coloured, flavoured, textured grain of life—could make sense of it, let alone find it worth striving after.

V

However, this is not the end of the story. I mentioned that in his commentary on Porphyry's introduction to Aristotle's *Categories* (philosopher's philosophy with a vengeance!) Boethius sets out a quite different ontology of forms; and it was this, rather than the Platonism of *De Trinitate* and *De Consolatione*, that so greatly influenced philosophers of the thirteenth century Aristotelian revival, such as Albert the Great and Thomas Aquinas.

Abstracting from details and differences, the general metaphysical view is that what makes an empirical object to be the thing it is, is not the *stuff* out of which it is made (*materia*) but its nature or *form* (*forma rei*). The pen which I am holding as I write this, is what it is as a metaphysically necessary consequence of having the *form* it possesses (i.e. its shape, construction and causal powers) and not in virtue of being a quantity of bakelite. This form is something we become aware of by studying the individual object. It is not something purely sensible, since, for example, function, and causal powers more generally, go beyond anything given in actual experience and warrant counterfactual claims. Still, by the exercise of reason and imagination in co-operation with sense-experience, we can say that in a broad sense the forms of things can be 'perceived'. This claim has two important aspects: first, the objects of such attention are individuals not universals; and second, they are objects of embodied experience not pure intellection.

It is not excluded by this view that there might also be the intellectual grasp of general natures abstracted from observation of individuals, but that is a further matter. According to one interpretation, what this involves is the intellect acquiring a nature identical to that of the abstracted form. Thus, for catness to exist as a universal nature apart from any actual cat's nature is just for some intellect to be informed by, and hence to be thinking of, the species-form catness. In the scholastic terminology of Aquinas, the actualization of the form is *one and the same thing as* the relevant actualization of the intellect thinking of it—*intelligibile in actu est intellectus in actu*.[15]

The relevant upshot is that a less extravagant version of Boethius' idea of philosophical consolation may now be available. We may say

[15] *Summa Theologiae*, I, q. 14, a. 2.

40

that human beings have the capacity to comprehend the natures of things, both individual and universal forms, and that the actualization of this capacity is a mode of self-realization through the exercise of our higher powers. Then, for reasons parallel to those given by Boethius, we may find it intelligible to say that this form of activity constitutes *a* purpose in life which renders it valuable and which eases or even eliminates 'the problem'. Assuming that the metaphysics of this view is intelligible and even plausible (and I believe it to be both), the questions remain of whether there is real reason to think that the contemplation of 'Aristotelian' forms is any more credible as philosophical consolation than Boethius' alternative, and whether there is any phenomenological basis for it. I shall end on what has to be a fairly brief consideration of these questions.

VI

Earlier, I mentioned the interpretation of a type of thoughtful experience associated with the observation of art and nature. The sort of thing I have in mind stands in need of philosophical investigation anyhow, and it would be a welcome outcome if the question of its nature and that of whether there is a form of philosophical consolation could be brought into complementary resolution. Here it will be best to proceed by example, and for reasons of personal familiarity but also because, as Aquinas says, visual perception is one of the most cognitive forms of experience,[16] I shall take my examples from painting and sculpture.

In the history of art an important growth point is marked by the career of Giotto—it is interesting and no accident that this was roughly contemporary with that of Aquinas.[17] In painting of the earlier Byzantine tradition subjects are treated in an apparently stylized iconographic fashion. This is not so much due to the loss of knowledge, formerly possessed by the Romans, of how to paint naturalistically, as to the neo-Platonic metaphysics that informed the Eastern Church Fathers who, in turn, shaped the theology of Christian culture. The Byzantine treatment of religious themes such as the crucifixion involves a timeless, 'viewpointless' presentation of the subject which is designed to suggest to the spectator the appearance of a transcendental realm. With Giotto, however, the depicted forms assume their natural embodiment. Figures possess the solidity and mobility of incarnate beings, are

---

[16] *Summa Theologiae*, I–II, q. 27, a. 1.

[17] For brief discussions of related matters see J. Haldane, 'Aquinas', Mediaeval and Renaissance Aesthetics' and 'Plotinus', in *A Companion to Aesthetics* D. Cooper (ed.), (Oxford: Blackwell, 1992).

seen to occupy an environment of earth and air and are caressed by a light whose 'temperature' is almost sensible.

So began a tradition of aesthetic naturalism that was developed in a range of ways over the next six hundred years. Within this history, though, is a particular strand of pictorial treatment of material forms that is difficult to describe but easy to recognize once one is attuned to it. As well as Giotto, painters in this 'tradition' include Masaccio, Fra Angelico, Piero Della Francesca, Bellini, Vermeer, Chardin and Morandi. The domination by Italian artists may be significant, since part of what defines this group is the quality of the light which passes around and between the objects depicted and also spreads over them. The light, the geometry of apparently simple forms and the softly glowing or crumbling colours create an atmosphere that beckons and reassures the viewer that therein all is for the best. The values realized are those of simplicity, humility, serenity, dignity and, one might now add in echo of Boethius, of sufficiency, reverence and goodness. Consider two responses to Chardin's still-lives:

> Here you are again, you great magician, with your silent compositions! How eloquently they speak to the artist! . . . how the air circulates among these objects, and sunlight itself does not better reduce the disparities between the things it falls upon. For you there are neither matching nor clashing colours . . .[18]

> . . . The miracle in Chardin's paintings is this: modelled in their own mass and shape, drawn with their own light, created so to speak from the soul of their colour, the objects seem to detach themselves from the canvas and become alive . . .[19]

These judgments are immediately intelligible and single out authentic features of the 'tradition', but they do not hit upon the particular, almost mystical quality that inheres in a few Chardins but is more clearly and more often present in the work of others of the group. The most telling of the later members is Morandi. Born in 1890, he was a near contemporary of de Chirico with whom he was an associate in the *Pittura Metaphysica* group of the last years of the first World War. But while the association with de Chirico under the 'metaphysical painting' description is intelligible it is also quite misleading. For when one considers relevant work by the two painters it is clear that the mysterious qualities in their canvases are quite different in each case.

What de Chirico achieves by his vacant neo-classical city-squares and shadow-hung cloisters is a sense of absence, as if the reality lies

[18] Diderot, Review of the Paris Salon of 1765, quoted in P. Rosenberg, *Chardin* (Lausanne: Skira, 1963), p. 96.

[19] E and J. de Goncourt, *Gazette des Beaux Arts*, 1863–64, in Rosenberg. op. cit. p. 101.

elsewhere and what we see is intended to question whether the normal course of experience is not itself a dream or a fantasy. In this he shares the challenging spirit of the surrealists. But apart from some early pieces that resemble de Chirico's works of the same period, Morandi's paintings have about them a quite different air of mystery. They do not challenge one's sense of reality but rather provide a kind of ontological reassurance. Things are what and how they are, and the contemplative grasp of this realizes a need to place oneself philosophically within the world. It is a great tribute to de Chirico, I think, that he recognized the special quality of Morandi's vision:

> Giorgio Morandi searches and creates in solitude . . . He looks at a collection of objects on a table with the same emotions that stirred the heart of the traveller in ancient Greece when he gazed on the woods and valleys and mountains reputed to be the dwelling places of the most beautiful and marvellous deities.
>
> These objects are dead for us because they are immobile. But he looks at them with belief, He finds comfort in their inner structure—*their eternal aspect.*
>
> In this way he has contributed to the lyricism of the last important movement in European art: *the metaphysics of the common object.* However much we may be aware that appearances deceive, we often look at familiar things with the eyes of one who *sees and does not know.*[20]

In conclusion, then, the thought which I wish to propose for further consideration is that there is a mode of thinking of the nature of things which is contemplative but which does not seek to transcend the realm of numerically distinct empirical forms. When it comprehends those forms for what they are, and *ipso facto* comprehends the immanent principles of being of individual objects, it is satisfied at having engaged with reality and thereby having realized itself. This is the consolation Boethius believed he had found, that of uniting oneself with the real, of coming to be at one with things—not, as mystics have often claimed, at one with everything, the totality itself being conceived of, in Parmenidean style as a unity, but united with each thing as one contemplates it for what it is.

Unlike Boethius' Platonic version, this account is believable and may be confirmed in the experience of a kind of art that defines itself in terms of the possibility of developing and communicating a sense of the being of things. If this account of the phenomenology of such works is

---

[20] G. de Chirico, 'Giorgio Morandi', from *La Fiorentina Primaverile* (1922), re-published in *Giorgio Morandi* (London: The Arts Council, 1970), p. 6.

correct, then the complementary resolution of which I spoke earlier has been achieved. The experience of a Morandi provides an entry into a perceptual mode of contemplative thought which constitutes one inter-pretation of the consolation of philosophy, and the consolation of philosophy answers the question: what explains the feeling one has, on seeing a Morandi, that in some sense all is well?

Let me add one final note by way of further evidence of a near equivalence between a certain kind of thoughtful art and a mode of philosophical reflection. In describing the relevant artistic tradition I spoke of pictorial treatments of objects, and I have thus far only discussed the work of painters. Among our contemporaries, however, are a number of sculptors who have sometimes achieved, and succeeded in communicating, the consoling vision of things as they are. Among these I would include Richard Long and Tony Cragg. While the characters and sensibilities of these artists differ markedly, at its best their work contributes a distinctive late-twentieth century element to the 'tradition'. Why that should be is suggested by the following remarks of each discussing their work. First Long:

> I like simple, practical, emotional
> quiet vigorous art.
> I like the simplicity of walking,
> the simplicity of stones.
> I like common materials, whatever is to hand,
> but especially stones, I like the idea that stones
> are what the world is made of.
> I like common means given the
> simple twist of art.
> I like sensibility without technique. (1980)

Then Cragg:

> The need to know both objectively and subjectively more about the subtle fragile relationships between us, objects, images and essential natural processes and conditions is becoming critical. It is very important to have first order experiences—seeing touching smelling, hearing—with objects/images *and to let that experience register.* (1982, p. 340)

Finally, the idea that a kind of spiritual enlightenment consists in experiencing the real for what it is, and being consoled by it, also finds eloquent expression in the following words put into the mouth of Kim by Rudyard Kipling:

> Then he looked upon the trees and the broad fields, with the thatched huts hidden among crops—looked with strange eyes unable to take up the size and proportion and use of things—stared for a still

half-hour. All that while he felt, though he could not put it into words, that his soul was out of gear with its surroundings—a cog wheel unconnected with any machinery . . .

He did not want to cry—had never felt less like crying in his life—but of a sudden easy, stupid tears trickled down his nose, and with an almost audible click he felt the wheels of his being lock up anew on the world without. Things that rode meaningless on the eyeball an instant before slid into proper proportion. Roads were meant to be walked upon, houses to be lived in, cattle to be driven, fields to be tilled, and men and women to be talked to. They were all real and true—solidly planted upon the feet—perfectly comprehensible—clay of his clay, neither more nor less. (1901)

In Boethius' figurative rendering, as *Philosophy* approached 'the clouds of my grief dissolved and I drank in the light'.

References

Belsey, A. 1991. 'Boethius and the Consolation of Philosophy, or How to be a Good Philosopher', *Ratio* (New Series), vol. IV.

Boethius. 1987. *The Consolation of Philosophy*, trans. V. E. Watts. (London: Penguin).

Chadwick, H. 1981. *Boethius: The Consolations of Music, Logic, Theology, and Philosophy*. (Oxford: Clarendon).

Cragg, T. 1982. 'Statement', *Documenta 7*.

Kipling, R. 1901. *Kim*. (London: Macmillan).

Long, R. 1980. *Words after the Fact*. (London: Anthony d'Offay).

Quine, W. V. 1981. 'Has Philosophy Lost Contact with People?', in *Theories and Things*. (Cambridge, Mass.: Belknap Press).

Stewart, H. and Rand, E. 1926. *The Theological Tracts*, Loeb Classical Library. *Boethius*. (London: Heinemann).

The real or the Real? Chardin or Rothko?[1]

ANTHONY O'HEAR

I

I will begin by considering some themes from Proust's wonderful essay on Chardin, *Chardin and Rembrandt* (Proust, 1988). Proust speaks of the young man 'of modest means and artistic taste', his imagination filled with the splendour of museums, of cathedrals, of mountains, of the sea, sitting at table at the end of lunch, nauseated at the 'traditional mundanity' of the unaesthetic spectacle before him: the last knife left lying on the half turned-back table cloth, next to the remains of an underdone and tasteless cutlet. He cannot wait to get up and leave, and if he cannot take a train to Holland or Italy, he will at least go to the Louvre to have sight of the palaces of Veronese, the princes of van Dyck and the harbours of Claude. Doing this will, of course, make his return to his home and its familiar surroundings seem yet more drab and exasperating.

> If I knew this young man I would not deter him from going to the Louvre, but rather accompany him there . . . I would make him stop . . . in front of the Chardins. And once he had been dazzled by this opulent depiction of what he had called mediocrity . . . I should say to him: Are you happy? Yet what have you seen but . . . dining or kitchen utensils, not the pretty ones, like Saxe chocolate-jars, but those you find most ugly, a shiny lid, pots of every shape and material (the salt-cellar, the strainer), the sights that repel you, dead fish lying on the table, and the sights that nauseate you, half-emptied glasses and too many full glasses. (1988, p. 123)

Proust then goes on to note that if one finds all this beautiful to look at, it is because Chardin found it beautiful to paint; and the underlying reason he and you find it all beautiful is because you have already unconsciously experienced the pleasure afforded by still life and mod-

[1] I intend, in this place, to develop further some of the argument of Oliver Soskice in his article 'Painting and the Absence of Grace', *Modern Painters*, Vol. 4, No. 1 (Spring 1991, pp. 63–5). I thank Michael McGhee for percipient comments on an earlier draft.

est lives, a pleasure Chardin had the power to summon to explicit recognition with his 'brilliant and imperative language'.

In sum:

> from Chardin we had learnt that a pear is as alive as a woman, that common crockery is as beautiful as a precious stone. The painter had proclaimed the divine equality of all things before the mind that contemplates them, before the light that beautifies them. (1988, p. 129)

He thus brings us out from a false ideal of conventional beauty, to a wider reality, in which in accordance with what Proust says, we are enabled to find beauty everywhere. Perhaps Proust is wrong about this: perhaps there are areas in which there is no beauty to be found, in a dying child's bootless agony, for example. But the underlying drift of what Proust says, and the substance of Chardin's aesthetic is surely right: that there is a real beauty in the midst of everyday domesticity, a beauty that may well be overlooked by aesthetic young men and other visionaries, who have a false ideal of beauty, one constrained by grandiosity and sublimity. Proust does not say this either, but one aspect of the false ideal of beauty is doubtless the tendency—so prevalent in the contemporary world—to treat the mundane as disposable; to fail to cherish it, to let it grow old and so become touched with humanity through use and familiarity; to fail to design it with care for its conformability to our sensibility, but to crush all that with a brash and ultimately impersonal dehumanizing aesthetic of function.

Another aspect of Chardin, to which I shall return, is the way in which in his still lives—in contrast to those of some of his brilliant Dutch predecessors, such as Kalf or Coorte—the objects emerge shyly, from a soft and often indeterminate background, against which they quiver in the light almost on the edge of visibility. Their being seen, and us seeing them is then represented by Chardin as what it is, a human achievement: the objects are summoned, as Proust puts it, 'out from the everlasting darkness in which they had been interred'. We may well be reminded by all this of Cézanne's words 'le paysage se reflète, s'humanise, se pense en moi'. These words, indeed, encapsulate what I want to emphasize in this article, particularly if they are considered alongside Cézanne's own mature *oeuvre*, which consists not of dead versions of life, so to speak, but of canvasses in which the appearances of things are vividly reconstituted for us out of the equally visible pigments of the paint (which is one reason why photographic reproductions are particularly faithless to the reality in the case of Cézanne).

In his essay, Proust goes on to contrast the reality Chardin evokes for us with what he calls the transcendence of reality in Rembrandt; in some of Rembrandt's works, objects become no more than the vehicles by which something else, another light, another meaning, is reflected. Whatever we might say about Rembrandt himself, it is possible to discern in western painting an oscillation of emphasis and interest between the everyday and the transcendent. Early Italian painters such as Duccio and Fra Angelico made little attempt to be fully realistic, being more concerned to give expression to the religious truths underlying the myths they painted. It was not that in their painting there was a complete inability to represent the appearances of things; they do and they can, but that is not the focus of their interest. In Raphael we find, to sublime effect, a balance of the physical and the religious, a balance which has been quite lost—though to stunning effect—in the opulent sensuality of Titian and Veronese. It is not for nothing that Titian has been seen as ushering in the materialism of the modern age, though to see him simply in such terms is to discount the pantheistic overtones and allegorical subject matter of paintings such as *The Flaying of Marsyas* and *The Death of Actaeon*. Given that painting is about the appearances of things, and that it is thus intimately related to our experience of seeing, it is even arguable that no painting can be regarded as materialistic in a reductionist sense; thus Monet—who is often taken to be the epitome of modernistic materialism in painting—in many paintings certainly emphasizes sensation at the expense of anything deeper or more inward, but his work also testifies to the intrinsic interest and value of our perspective on the world, and would thus resist analysis of that perspective as 'mere' epiphenomenon.

If Monet was not interested in his art in anything specifically religious, there were other painters in the nineteenth century who were, and who, for a time, managed to combine a stylistic naturalism with an overtly religious content. We can think here of the Nazarenes in Germany and the Pre-Raphaelites in England. But this affected naïvety could not last. As Peter Fuller graphically described in *Theoria* (1988) under the assault of Darwinism, the hope expressed in *The Light of the World* quickly turned to the desolation of Holman Hunt's *Scapegoat* and Dyce's *Pegwell Bay*. In the former, a visit to the Holy Land failed to elicit anything more uplifting than a mangy animal and a waterless, wasted landscape. In the latter, women and children hunt for fossils on an empty beach beneath a sickly sky across which Donati's comet is passing. The symbolism would in both cases have had a direct impact on the intended audience, even if it needs conscious retrieval on our part. Moreover, if the natural world could no longer be seen as directly revelatory of the divine hand, what point was there in its literal depiction, particularly when the camera could do that painlessly? At least some of the Pre-Raphaelite Brotherhood retreated into highly-charged medieval fantasy (and not *just*, I surmise, because Rossetti was no natural draughtsman).

Other religiously motivated artists worked in more oblique ways than the Pre-Raphaelites. Caspar David Friedrich's romantic landscapes turn out to be carefully crafted allegories of the Christian's journey through the world, but in them there is little enough sense either of the divine nature of the world—which is all too often a hostile environment redeemed only by the distant presence of a cross or of a vision of a cathedral—or of the concrete detail of the Christian myth.

Another approach is that of van Gogh: to transform the natural in a visionary manner: to see stars, olive trees, cypresses and the like, not as they appear in everyday life but as transfigured symbols of a deeper more vibrant life pulsating beneath the empirical surface, and visible to the man of faith. Van Gogh and Friedrich, whatever their differences in approach and in painterly quality, concur in their desire to preserve what they would see as the essence of the Christian message and in their refusal in the main either to represent that message literally or to see the natural world as it is to the normal eye as a straightforward manifestation of the divine. And in all these respects, they have been followed by many artists in this century.

With the decline of natural theology and the collapse of credal religion, the problem for a religiously motivated artist is to devise a way of presenting the essence of religion without illegitimate recourse to either creed or nature. It is hardly surprising that some artists should have sought abstraction as a way forward, nor indeed that some critics and commentators have seen Barnett Newman and Mark Rothko as among the supreme religious painters of our time. Newman indeed told us that we should take him in this way, providing elaborate references to Kabbalistic themes in his titles and commentaries. I must confess to some difficulty in accepting Newman on his own terms. His huge rectangular expanses of flat colour punctuated by vertical stripes are not easily experienced as pointers to the numinous, although the canvasses can, like those of Rothko, engulf the perceiver by their sheer size, and one can be amazed for a time at Newman's sheer effrontery.

By comparison with Newman, Rothko's major canvasses are not so big (a mere 2 metres wide in many cases), but they all produce an experience of engulfing the perceiver, as much by their working from a ragged, indeterminate edge to quasi-rectangular expanses of deep colour, as by their size. They are, in a way, perfect expressions of the world as a stage on which everything is about to happen, or has already happened. In Rothko's work, there is no trace of the concrete, nothing appears; we are overwhelmed by hazy, empty sublimity. And before long, I find them deeply unsatisfying, longing to turn to the modesty and concreteness of a Chardin. Is engulfment, the wiping away of all determinations and horizons, what life—and art—is all about? If it is, then human effort and perception and perspective are, in the final analysis, mocked. There is, in fact, more than a grain of truth in Patrick Heron's barbed comment: 'that having painted 800 such canvasses, Rothko was led nowhere, but to the dealers and suicide!' (1989, p. 39)

II

My excursion into the history of painting is not simple self-indulgence. I have engaged in it in order to illustrate as vividly as I can the reason why I am unable to rest content with approaches to existence and experience which would undervalue human perception and human experience.

Either there is some cosmic point to human existence, or there is not. In either case, the value of human experience remains irreducible, despite temptations on both accounts to discount it.

Let us suppose, first, that human beings are simply products of a mindless, purposeless cosmos, thrown up by the random or mechanistic activities of more basic particles. Then it will be true that our perspective on the world (including our perceptions, our feelings and our meaning) is itself a by-product of more fundamental processes and reflective of no deeper reality or purpose. It may even be that knowledge of these fundamental processes and the laws which govern them would enable us to predict human perceptions and actions. Relative to the more fundamental processes which underlie our perceptions, the modalities of our perceptions (colour, taste, smell, touch, sound) will be regarded as consisting of secondary qualities, qualities which arise only as a result of the interaction of colourless, tasteless, odourless, textureless and silent particles with our sense organs. But even if this were true, our perceptions, and artistic works devoted to the exploration and development of human perspectives would not lose any value they have for us. The value of a Monet landscape lies not in its genesis, but in the satisfactions and delights and insights it affords us, satisfactions and delights and insights which would not be corroded completely even were we to adopt the scientific view from nowhere, that which displaces the human subject from the centre of things, and sees human life and perception as part of more inclusive causal processes. But even accepting a story of this sort at an intellectual level, we still feel and experience things as we do, and it is in our lives and experiences as lived and experienced that value lies.

Indeed, even seeing our *Lebenswelt* as the product of primary material processes does not make what is revealed in our world *false*. It is open to us, even while accepting the scientific view as causally fundamental, to regard what is revealed in our experience as a legitimate disclosure of the world, one which is available only to us. Just

51

because the world of so-called secondary qualities arises only in the interaction of particles with our sense organs, does not mean that that world is unreal or in some derogatory sense subjective, any more than the pictures a television set emits are unreal or subjective just because a television receiver has to transform invisible radio waves into visible images.

The idea that there is in human perception a singular and irreducible revelation of the real world becomes even more plausible if we see human existence as cosmically intended in some way. For what, from the cosmic point of view, could be the point of human existence, other than that the cosmos should be experienced and understood in a human way? It is worth noting here that not even God, being a-temporal, a-spatial and immaterial, could know what it is like to be a human being; however much God might have foreknowledge of our thoughts, experiences and actions, this foreknowledge would necessarily be schematic, abstract and theoretical.

It has for long seemed to me, as it did to Oliver Soskice in 'Painting and the Absence of Grace', that if we are here for a purpose, and if human life has something unique to contribute to the cosmos, it is this:

Sind wir vielleicht *hier*, um zu sagen: Haus,
Brücke, Brunnen, Tor, Krug, Obstbaum, Fenster—
höchstens: Säule, Turm . . . aber zu *sagen*, verstehs,
oh zu sagen *so*, wie selber die Dinge niemals
innig meinten zu sein.

'Are we perhaps here to say: House,
Bridge, Well, Gate, Jug, Fruit Tree, Window
—at most, Pillar, Tower . . . but to say—oh!
to say in this way, as the things themselves
never so intently meant to be.'
(R. M. Rilke, *Ninth Duino Elegy*, lines 32–6)

That is, we are here to experience and articulate something about things, something which things themselves can neither articulate nor experience, but which also (as Rilke goes on to say) is beyond the power of angels to know and experience.

We can, of course, see value in doing what Rilke says we are here to do, even if we are not put here to do anything, and that valuing need not be impugned by scientific explanations or views from nowhere. It is, perhaps, strange then that religion, which does see us as being on earth for a reason, all too often downgrades our Rilkean task, for doing what Rilke says we are here to do is the one thing we alone can do; and is the reason I place a higher value on the aesthetic experiencing of the world than many theological writers. I see the contemplation of what arises in human practice as the singular contribution we as humans can make to the cosmos. That is to say, through our practices and the associated sensory apparatus, we divide the world up in various ways, but because we are self-conscious we can reflect on enjoying these perceptions and evaluate the significance of what is revealed in our practices and perceptions. Because of our status as sensory *and* intellectual, we alone are in a position to enjoy particular perceptions of the world, and to evaluate the fruits of those perceptions. A merely sensory consciousness could not reflect on what it perceives, while a purely intellectual being (an angel) would perceive or experience nothing.

Religion, though, would take us all too quickly from the human to something we cannot envisage or articulate at all. In so doing, it all too easily downgrades and wipes away the human. Soskice refers in his article to Hölderlin's *Griechenland*:

> . . . where the longing for eternity knows no bounds
> Divine things are overcome with sleep.
> There is no trust in God, no proportion . . .

I find more than a trace of this lack of proportion in John Hick's *An Interpretation of Religion* (1989). I am now going to turn to this book as it is, I believe, a brave and radical attempt to salvage something of religious value from the downfall of dogmatic religion; or, more precisely, from the dilemma which arises for any would-be religious believer from the existence of a plurality of religious faiths, all of which seem to have some good claim to be regarded as offering genuine insights into the divine.

Hick is also faced with the problem that straightforward dogmatic religion is hardly credible in the late twentieth century. For him, there is to be no return to a Pre-Raphaelitic naïvety. Moreover, he is well aware that Christianity is not the only credal contender. Right at the start of his book he speaks of the transcendent being perceived through different and distinctive cultural lenses (1989, p. 8).

Nevertheless, he is convinced that there is a Real behind the different religious traditions: that is to say, that the God of Abraham, Isaac and Jacob, the Holy Trinity, Allah, Vishnu, Brahman, the Dharmakāya/Nirvāṇa/Śūnyatā, Zen and the Tao, all represent ways of affirming the same ultimate.

I presume that Zeus, Jupiter, Wotan, and the other limited, personal deities are not included in Hick's list because they are limited; they are themselves subject to fate and to contract and are not transcendent. Be that as it may, and leaving aside the point that Jahweh, Jesus and Allah are all conceived in personal—and hence determinate and limited—terms, I want now to consider the implications for us of conceiving what he calls the real as Hick does, as a metaphysical or noumenal ground underlying all religious objects, whether personal (as often in

Western tradition) or impersonal (as characteristically in eastern traditions): that is, something divine and real behind the various humanly mediated revelations or intimations of divinity.

There is, indeed, as Hick points out, a drive in human thought to seek the utterly transcendent or the self-subsistent ground of our beliefs and valuations. We are not just conscious, we are also self-conscious. We have beliefs and values, and we are aware of having them. I am aware that any belief or judgment I make is mine, and that my perspective is just that—*my* perspective on a world which has an existence independent of me, and in which there are other agents also making judgments of value. I thus become aware that my belief or judgment is not the only possible belief or way of according value. I become aware of my route through the world as only one of many possible routes, and thereby open to question.

My self-consciousness may well be sparked into reflective activity by the realization that there are people other than me and cultures other than mine. But once activated, we begin to realize the limited nature of any actual sets of beliefs or values we have. We thus formulate for ourselves the conception of an absolute truth, an absolute good. But in formulating such notions, we realize that no merely human sets of beliefs or values can be guaranteed to be absolute: all will be more or less limited by the particular perspective we adopt.

Hick sees a great upsurge of self-conscious dissatisfaction with local and particular beliefs and customs around the fifth century BC, the time of Confucius, Lao Tzu, the Buddha, Mahavira, Zoroaster, the Hebrew prophets—and Socrates. As he puts it, following Karl Jaspers, at what is dubbed, the 'axial' age,

[I]ndividuals were emerging into self-consciousness out of the closely-knit communal mentality of their society. . . . Religious value no longer resided in total identification with the group but began to take the form of personal openness to transcendence. And since the new religious messages of the axial age were addressed to individuals as such, rather than as cells in a social organism, these messages were in principle universal in their scope. (1989, p. 30)

This 'post-axial' quest for individuality and universality is in fact based in our very nature as self-conscious beings, as I have been arguing, and is not just a product of specific historical circumstances, however much some circumstances might encourage its development. The drive to individualism and the search for universal, unlimited truth and value are inherent in human nature, and cannot be totally suppressed. Nevertheless, we should realize that both these tendencies carried with them two, probably connected problems. First, they destroy culture and secondly they encourage a religion of unknowing, whose effect may well be to undermine the human.

In saying that individualistic and universalistic attitudes destroy culture, what I have in mind is the complaint raised by Aristophanes (in *The Frogs*) and by Nietzsche (in *The Birth of Tragedy*) against Socrates. In their view, the greatness of ancient Athens stemmed in part from the fact that its citizens were united in reverence for a myth or set of myths which bound them together (and which also, doubtless, enabled them to make culturally crucial distinctions between the best and the rest of society, between themselves as Athenians and other Hellenes, and between Hellenes and barbarians). But Socrates and, to an extent, Euripides, taught ordinary men to question the myths and to cease to respect their superiors or to regard Greeks as Greeks first and as men second.

In *The Decline of the West* (1926), Spengler characterized the transition from what he called Culture to Civilization in the following terms, which certainly have a bearing on Hick's characterization of axiality:

> In place of a type-true people, born of and grown on the soil, there is a new sort of nomad, cohering unstably in fluid masses, the parasitical city-dweller, traditionless, utterly matter-of-fact, religionless, clever, unfruitful, deeply contemptuous of the countryman, and especially that highest form of countryman, the country gentleman. (1926, p. 32)

And Spengler goes on to speak of the 'uncomprehending hostility' of the new city-dweller to 'all the traditions representative of the Culture (nobility, church, privileges, dynasties, conventions in art and limits of knowledge in science)', of his 'keen and cold intelligence that confounds the wisdom of the peasant' and of his apparently new-fashioned but actually quite primitive and instinctual naturalism in all matters of sex and society.

While there will in city-dwelling civilizations tend to be a decline of local and particular ways of doing things, in favour of the universal, the reproducible, the purely functional and the disposable, intellectually the decline of (Spenglerian) culture presages an attempt to discern a unity underlying apparently disparate forms of similar activity. This, of course, is what Hick attempts in the case of religion, and whose aesthetic analogue is the *œuvre* of Rothko. I say this in the case of Rothko because I take it that what he is presenting is not just colour as an end in itself, but colour as a symbol of the ineffable.[2]

---

[2] This might raise the question as to what I would say about a work—an Ellsworth Kelly, say—in which the aim might well be taken to be just the presentation of an experience of colour, as it is in itself and for itself, in which a painting 'stands', as a large and public sense-datum. Leaving aside the gigantism such works are typically prone to, I would have to say that this is a further

The attempt to fuse what, on the face of it, is unfusable (e.g. Islam, Christianity, Hinduism and Buddhism) together with the religious drive to find an underlying ultimate reality conspire to produce a religion which is, practically speaking, without content. Drawing, indeed, on ancient religious texts and traditions, Hick speaks of God, Brahman, the Dharmakāya as unlimited, 'not to be equated without remainder with anything that can be humanly experienced and defined' (1988, p. 236); the Real in itself cannot be said to be 'one or many, person or thing, substance or process, good or evil, purposive or non-purposive' (1988, p. 246); 'we postulate the real *an sich* as the ultimate ground of the intentional objects of the different forms of religious thought-and-experience' (1989, p. 350). Even if it is said that there is *something* (the Real) which underlies all the phenomenal divinities, the fact is that we can say nothing about this real. It is hard to see the difference between faith in a non-describable Real and agnosticism.

I hope that by now the point of my earlier reference to the work of Mark Rothko is now clear. Instead of anything specific we are, in both cases, being offered a void, an emptiness, which is said to be pregnant with noumenal meaning and to underlie the merely phenomenal. Hick does urge us to respect and to maintain the disparate phenomenal manifestations of the Real (i.e. the actual world-religions). But given the superior viewpoint he is urging us to adopt, whereby each of these religions is seen as a radically incomplete version of something utterly ungraspable, of which we are told there are other equally valid (though I would add) mutually inconsistent manifestations, it is hard to see how one could in all good conscience continue to worship in, say, a Christian church or a Muslim mosque.

Even if the exclusivist and particular claims of the actual world-religions could somehow be mutually reconciled against the background of a noumenal we know not what, I am extremely dubious about the moral and human effects of worshipping and directing our efforts to a Being as indeterminate for us as Hick's unknowable Real. If this Real can be said to be neither person nor thing, good or evil, purposive or non-purposive, loving or hating, as Hick avers (1989, p. 350), what ultimate reason is there for us to love, to be good, to respect others or to engage in purposive activity at all? Hick does indeed struggle manfully

---

area in which I would distance myself from a Proustian equality of all things; that there is more meaning, more humanity, in a wineglass or a firedog than in a patch of yellow, however large or small—and that the effort of a painter to render the one is potentially more worthwhile from a human point of view than the effort to represent the other. We value Vermeer's little patch of yellow but we do not value it just because it is what it is in itself. We value it because it is part of an extraordinarily gentle and precarious humanization of the world.

with the Buddhist śūnyatā (emptiness, transcendence of all perspective (cf. 1989, pp. 288–92)); śūnyatā seems to be the natural end of life if the Real is as Hick conceives it. At any rate, it is hard to value human activity or to see how our way of perceiving things could be a worthwhile revelation of a Reality which is essentially unknowable.

Hick doubtless would say that there is a reason for us to cultivate loving and truthful attitudes, given that we are aiming to eradicate our ignorance and delusion with respect to the Real, and also that such attitudes are propagated by all the great world religions. But I have to say that from Hick's perspective of radical agnosticism, even loving and truthful attitudes on our part will get us no closer to the Real: we will still have nothing to say about it, and nothing to grasp, except possibly by some incommunicable, and hence dumb, religious experience. I can, indeed, see nothing in Hick's account to rule out a Real whose ultimate nature was not, say, closer to a Nietzschean will-to-power than to a Catholic Sacred Heart; nor indeed am I convinced that compassion for suffering humanity is always to the forefront of the world religions, as opposed to a chronic carelessness about individual life and suffering in face of cosmic dramas of global, rather than individual redemption or transformation. What is there in the notion of the unknowable, transcendent Real to rule but the possibility that our idea of a compassionate divinity is simply the ultimate fantasy of a deluded humanity whose final fate is to be broken on the wheel of existence? And I do not think it can be denied that some religious seekers after an ultimate divine reality have found not bliss but an emptiness, even a cruelty, too terrible to contemplate.[3]

Against such a background the emptiness—at its worst, the rhetoric—of Rothko would be vindicated against the painstaking and human modesty of Chardin, and what Chardin presents to us as an all-too-fragile achievement will be swallowed up in the abyss of the divine. At the same time, it is doubtless true that we come to see Chardin's achievement as the achievement it is just when we begin to understand that we are standing above an abyss, cosmically speaking, and that

[3] In the light of that dark night of the soul which is a recurring theme throughout the history of religious practice, I have to say that I find the efforts made by Hick in his book (1989, pp. 304–6) to link religious practice to 'politico-economic liberation' (i.e. anti-capitalism) sentimental and misguided religiously as well as economically. His endorsement of the 'basic intent' of Marx, Lenin, Trotsky and Mao, as a 'dispositional response of the modern sociologically conditioned consciousness to the real' (p. 306) serves, I think, to underline the extent to which religious thinkers, of all ages, get embroiled in the delusions of their time. Invoking a Real underlying the delusion will be of little consolation to those whose lives, domesticity and all, are ruined by projects of politico-economic liberation.

Anthony O'Hear

human domesticity and human perception rest on no secure foundation. In terms of my illustrative analogy, then, Rothko's Real might be seen to serve as the background from which Chardin's reality—and ours—emerges and is perceived.

References

Fuller, P. 1988. *Theoria*. (London: Chatto and Windus).
Heron, P. 1989. 'Can Mark Rothko's Work Survive?', *Modern Painters*, Vol. 2, No. 2 (Summer) pp. 36–9.
Hick, J. 1989. *An Interpretation of Religion* (London: Macmillan).
Proust, M. 1988. *Against Sainte-Beuve and Other Essays*, trans. J. Sturrock. (Harmondsworth: Penguin Books).
Spengler, O. 1926. *The Decline of the West*, trans. C. F. Atkinson. (New York: Barnes & Noble).

Love and Attention

JANET MARTIN SOSKICE

I

The matched pair 'love' and 'attention' is familiar to most of us from the essays in Iris Murdoch's *The Sovereignty of Good*. Although she tells us in that book that there is, in her view, no God in the traditional sense of that term, she provides accounts of art, prayer and morality that are religious. 'Morality', she tells us, 'has always been connected with religion and religion with mysticism' (Murdoch, 1970, p. 74). The connection here is love and attention: 'Virtue is *au fond* the same in the artist as in the good man in that it is a selfless attention to nature' (ibid. p. 41). Art and morals are two aspects of the same struggle; both involve attending, a task of attention which goes on all the time, efforts of imagination which are important cumulatively (p. 43). 'Prayer', she says, 'is properly not petition, but simply an attention to God which is a form of love' (ibid. p. 55).

Murdoch freely acknowledges her indebtedness to Simone Weil and the writings of both, in turn, have influenced many others—amongst whom, recently, is Charles Taylor in his book *Sources of the Self*. For Taylor, too, moral and spiritual intuitions go together. We must ask what we love, what we attend to, in order to know who we are and what we should be. He insists that 'orientation to the good is not . . . something we can engage in or abstain from at will, but a condition of our being selves with an identity' (Taylor, 1989, p. 68).

So it seems that 'love' and 'attention' mark a place of confluence for the concerns to which the essays in this volume address themselves, Philosophy, Religion and the Spiritual Life, with the further desirable feature that, as they have been discussed and as I shall discuss them, the ethical too is central.

Weil, Murdoch and Taylor, but especially the two last, draw similar portraits of that which they admire and that which they eschew. Love is a central concept in morals. To be fully human and moral is to respond to that which demands or compels our response—the other attended to with love. It is this loving which both draws us out of ourselves and which constitutes us fully as selves. For Murdoch the best exemplar of the 'unselfing' by attention is our experience of beauty.

I am looking out of my window in an anxious and resentful state of mind . . . suddenly I observe a hovering kestrel. In a moment everything is altered. The brooding self with its hurt vanity has disappeared. There is now nothing but kestrel. (Murdoch, 1970, p. 84)

We respond to the 'Good' or the 'Beautiful'. There is a debt to Plato in the idea of a Good, the love of which empowers us to do good and be good. As Taylor says, this 'takes us far beyond the purview of the morals of obligatory action' with which much modern mainstream moral theory has contented itself. (Taylor, 1989, p. 93). Ancient philosophical accounts of practical reason both Platonic and Aristotelian, Taylor argues, were substantive and implied that 'practical wisdom is a matter of *seeing* an order which in some sense is in nature' (ibid. p. 86). And in the works of Weil, Murdoch and Taylor metaphors of vision and seeing are deliberately employed. Murdoch uses 'attention' 'to express the idea of a just and loving gaze directed upon an individual reality'. This she believes to be 'the characteristic and proper mark of the active moral agent' (Murdoch, 1970, p. 34). Attentive love is close to contemplation.

The indebtedness of this line of thought to Christian spirituality, as well as to ancient philosophy, is evident and acknowledged. But any note of self-congratulation on the part of the Christian must be short-lived when one realizes that what is criticized in these theories is also recognizably the product of a Christian tradition of spirituality. Neither Murdoch nor Taylor have much time for the 'man of reason' who in various guises trudges through the works of early modern philosophy, a disengaged self in the disenchanted universe. Although readers may be familiar, I cannot forbear quoting Murdoch's description of Rational Man,

How recognizable, how familiar to us, is the man so beautifully portrayed in the *Grundlegung*, who confronted even with Christ turns away to consider the judgement of his own conscience and to hear the voice of his own reason . . . this man is with us still, free, independent, lonely, powerful, rational, responsible, brave, the hero of so many novels and books of moral philosophy . . . He is the offspring of the age of science, confidently rational and yet increasingly aware of his alienation from the material universe which his discoveries reveal. (Murdoch, 1970, p. 80)

This is he who, in Taylor's words, is 'capable of objectifying not only the surrounding world but also his own emotions and inclinations, fears and compulsions, and achieving thereby a kind of distance and self-possession which allows him to act "rationally"' (Taylor, 1989, p. 21).

This new agent of science gains control, even in his moral life, through 'disengagement' and objectification. Indeed Taylor argues it is only through a disengagement effected by radical subjectivity that the new radical objectivity is possible. Once confined securely within our selves we can manipulate and control a world of objects (Taylor, 1989, pp. 173–4). Even our affective responses come to have a value analogous in early modern philosophy and science to secondary properties such as 'red' and 'pain'. Accurate knowledge asks us to 'suspend the 'intentional' dimension of experience, that is, what makes it an experience *of* something' (ibid. p. 162).

Taylor underlines the Augustinian ancestry of this disengaged self, and its radical reflexivity. Indeed this miracle of self-mastery is a familiar figure in the texts of spiritual theology. I am tempted to say that despite the criticisms the 'disengaged self' or 'Rational Man' has received in recent years from philosophers, his theological near-relation, Spiritual Man, has continued virtually unchallenged, especially in the area I shall call 'received spirituality'.

II

For each of us, no doubt, a vision is conjured by the phrase, 'the spiritual life' and for most, I'd wager, that in our personal lives at least this is an eschatological vision—something piously hoped for in the future but far from our daily lives where, spiritually, we just 'bump along'. I believe we can also speak of a 'received view' of spiritual life which in its Catholic Christian form might involve long periods of quiet, focused reflections, dark churches and dignified liturgies. In its higher reaches it involves time spent in contemplative prayer, guided or solitary retreats, and sometimes the painful wrestlings with God so beautifully portrayed by the Metaphysical Poets. Above all it involves solitude and collectedness. It does not involve looking after small children.

I have been in the past envious and in awe of colleagues (usually bachelors) who spend their holidays living with monks in the Egyptian desert or making long retreats on Mount Athos. They return refreshed and renewed and say such things as 'It was wonderful. I was able to reread the whole of *The City of God* in the Latin . . . something I've not done for three or four years now.' I then recall my own 'holiday' as entirely taken up with explaining why you can't swim in the river with an infected ear, why two ice creams before lunch is a bad idea, with trips to disgusting public conveniences with children who are 'desperate', with washing grubby clothes, pouring cooling drinks, and cooking large meals in inconvenient kitchens for children made cranky by too much sun and water. From such holidays one returns exhausted and wondering why people go on holidays. But middle-class family holidays are only memorable instances of a wider whole. Parents of small children find themselves looking enviously over the wall at their more spiritual brethren—are these not the true 'spiritual athletes' whose disciplined life of prayer brings them daily closer to God?

The 'received view' of the spiritual life seems to confirm this, as does a good deal of guidance from priests and pastors. One story will have to suffice. A devout Anglican woman of my acquaintance had her first baby. Like most new mothers she was exhausted, but she was also distraught to find her devotional life in ruins. She took advice from three priests. The first told her that if the baby woke at 6.00 a.m., she should rise at 5.00 a.m. for a quiet hour of prayer. The second asked if her husband could not arrange to come home early from work three times a week so that she could get to a Mass. This advice proved threatening to life and marriage. The third told her, 'Relax and just look after your baby. The rest of the Church is praying for you.'

The advice of the third was the best and shows, too, why one does not really resent the retreatant on Mount Athos, or the religious contemplatives. These people, on the Catholic model at least, are praying for us all. But still the priest's advice is not entirely consoling. Is the busy new mother a sort of Christian 'on idle'? Will others carry on seeking God's face while you spend six or eight or twelve years distracted by the cares of the home? Is this the 'Martha' phase of life when you run the creche and make the tea, while the real work of attending to God is elsewhere? Not surprisingly many new mothers feel slightly bitter about this state of affairs.

Despite markers that could lead elsewhere, Christian 'received spirituality' is still shaped by particular views of contemplation, contemplative prayer and religious ecstasy that disenfranchise many people, and perhaps especially women.[1]

The 'received view' has a noble ancestry. Consider Gregory of Nyssa's influential treatise 'On Virginity', written sometime around A.D. 368. It is not easy, says the author, to find quiet for Divine contemplation within secular life and, as he would create in his readers a passion for excellence he recommends as a 'necessary door of entrance to the holier life, the calling of Virginity' (Nyssa, 1979, p. 343). His praise of virginity takes an interesting tack. He does not, as might be expected by our prurient age, condemn sexual activity. Rather he reserves his disapprobation for marriage, even for an 'ideal' marriage. Consider a marriage in every way most happy—illustrious birth, competent means, deep affection. Beneath these blessings 'the fire of an inevitable pain is smouldering' (ibid. p. 345). The young wife will grow old and die, she may on the other hand die young in childbirth, and the

---

[1] 'Received spirituality' is probably also a fantasy remote from the actual busy lives of many monks and nuns.

child with her. Children born safely may be subject to accident, illness and disease. You (male) may die on a business trip. A young wife is soon a widow, friends desert her, families quarrel, finances fall to ruin. In short, family life is one damn thing after another. 'He whose life is contained in himself . . .', says Gregory, can easily bear these things, 'possessing a collected mind which is not distracted from itself; while he who shares himself with wife and child' is totally taken up with anxiety for his dear ones (ibid. p. 347).

The striking thing about Gregory's analysis is that it is so convincing. He is simply right, and while we in the affluent west may be spared many terrors of deaths in childbirth, we have no difficulty enumerating other vexations which erode time and energy and would take us from contemplative quiet in the way Gregory describes.

But what about the medicine he prescribes? There is only one way, he says, to escape from Nature's inevitable snares and

> it is, to be attached to none of these things, and to get as far away as possible from the society of this emotional and sensual world; or rather, for *a man to go outside the feelings which his own body gives rise to*. Then, as he does not live for the flesh, he will not be subject to the troubles of the flesh. (ibid. pp. 350–1, my italics)

He will not be disturbed then by the troubles of his own flesh, nor by the disturbing and demanding flesh of spouse and children. By this means, Gregory says, we may emulate the spirits who neither marry nor are given in marriage but rather 'contemplate the Father of all purity' (ibid. p. 351).

> How can the soul which is riveted to the pleasures of the flesh and busied with *merely human longings* turn a disengaged eye upon its kindred intellectual light? . . . The eyes of swine, turning naturally downwards, have no glimpse of the wonders of the sky; no more can the soul whose body drags it down look anymore upon the beauty above; it must pore perforce upon things which though natural are low and animal. To look with a free devoted gaze upon heavenly delights, the soul . . . will transfer all its power of affection from material objects to the intellectual contemplation of immaterial beauty. (ibid. p. 351, my italics)

Once freed, the soul in its virgin state can emulate the God who is pure, free and changeless. Gregory takes seriously the idea that man is made in the image of God, but transcribes from an idealized 'Man' a picture of God as sovereign, rational and free, the very image of 'disengaged man'.

Even if we allow a little space for rhetorical excess it cannot be doubted that Gregory's Treatise invokes a spiritual ideal in which the demands of others, even of one's own babies and children, are not merely indifferent to the task of gazing on God, but in competition with it. The higher life is akin to that of Plato; reason, defined in terms of a vision of order, purity and immutability, governs desire. The 'good' man is 'master of himself' (Taylor, 1989, p. 115).

A distinguished Latin counter-part to Gregory's essay may be found in the first book of *De Doctrina Christiana* where Augustine develops his famous distinction between things which we are to enjoy and things which we are to use. That which we enjoy makes us happy, we rest with satisfaction in it for its own sake. Those things which are objects of our use, on the other hand, help us attain to that which makes us happy. But should we set ourselves to 'enjoy' what should properly be 'used' we are hindered on our way.

Augustine illustrates this with a favoured simile of the voyage: suppose we were wanderers in a strange country and could not live happily away from home. We must use some mode of conveyance to return. But the beauty of the country through which we pass or the pleasure of travel may divert us from 'that home whose delights would make us truly happy. Such is a picture of our condition of this life of mortality. We have wandered far from God; and if we wish to return to our Father's home, this world must be used, not enjoyed . . .' (1988, p. 523). It is our duty rather to 'enjoy the truth which lives unchangeably', for no one, according to Augustine, 'is so egregiously silly' as to doubt 'that a life of unchangeable wisdom is preferable to one of change' (ibid. p. 525). It is only the strength of evil habits that draws us to less valuable objects in preference to the more worthy. Human loves, as Augustine knew, bring bereavement and sorrow; 'those only are the true objects of enjoyment' which are 'eternal and unchangeable. The rest are for use . . .' (ibid. p. 527). Even our neighbour, whom we are commanded to love, we love for the sake of something else—that is, in Augustine's terms, 'we use him'. The contrasts are between the eternal, changeless and Divine and the temporary and material. The latter— even one's own children—should only be used on our way to the former.

Few of us are likely to be attracted or convinced by Augustine's account of 'enjoyment' and 'use', an account which even he may have regarded as 'experimental and finally inconclusive' (O'Donovan, 1980, p. 26). Neither Gregory nor Augustine was ultimately successful in stopping Christians from marrying and forming attachments to husbands, wives, and children. What they, or a complex tradition devolving directly and indirectly from them, may have been more successful in introducing is a particular idea of the 'spiritual life' still much present in 'received spirituality'. For there emerges between those wallowing in the vexations of secular life and the vision of God in which the blessed share, a distinctive account of an intermediate position of those who are *in via*. It is at this point that we may more readily be convinced by the Augustinian picture, for while none in this life is likely to reach the 'homeland', serious sojourners in the spiritual life may nonetheless establish themselves on the way. A hierarchy is established which privileges the detached life over that of affection and disruption and it is no coincidence that this spiritual hierarchy can be mapped onto other orderings. It is not simply a contrast between the cloistered life and the secular but is aligned with the distinction, common to ancient philosophy, which contrasts the demands and turmoil of ordinary domestic life with the excellences of the life of the polis—the life according to reason, the life of the philosopher, the citizen, and the lover of beauty. Such distinctions, in classical antiquity, ran along overt lines of sex and class. Women, children and slaves, as inhabitants of the rational demimonde, pursue life's necessities. Adult, male, free citizens pursue what Taylor calls the 'good life', 'men deliberate about moral excellence, they contemplate the order of things . . . decide how to shape and apply the laws' (Taylor, 1989, 211–12). Cloister, academy and law court are judged more suited to the true ends and excellences of human beings than are kitchen and nursery.

The contrast of 'ordinary life' and higher calling is not without its philosophical representatives today—one case seems to be Hannah Arendt's distinction in *The Human Condition* between productive, artifact-generating *work* and repetitious, inconclusive *labour*. Most of what women and slaves have done is of course the latter—the endless cycle of making meals which will only be eaten and washing clothes which will only be soiled. But even our advocates of love and attention seem sometimes to prefer to illustrate their thesis with relatively fixed or 'pure' objects. For instance Iris Murdoch's preference for beauty as the best evidence for a transcendent principle of the good. Ordinary human love, she says, is normally 'too profoundly possessive and also too 'mechanical' to be a place of vision' (Murdoch, 1970, p. 75).

It would be rash and inaccurate to suggest that exaltation of the spiritual life (so fashioned) has always in Christian history meant the denigration of family life (though sometimes it has).[2] You will be quick to point out the many places at which theologians and poets (even

---

[2] It has been pointed out to me that I have paid no attention in this paper to the Lutheran and Reformed traditions of domestic holiness. This is largely deliberate, since I'm addressing a version of 'spiritual life' associated with the catholic aspects of Christianity. Indeed the phrase 'spiritual life' is one which Catholic Christians (a category which includes more than Roman Catholics) are more likely to use than are Protestants. However, it is also the case that women friends of mine in the Reformed tradition tell me that, despite the positive signs, things are not much better for them in their own churches.

Metaphysical poets) have praised the daily round and trivial task. But for the most part such things as attending to a squalling baby are seen as honourable duties, consonant with God's purposes, rather than in themselves spiritually edifying. Most Christian women, for instance, think that what they do around the home is worthy in God's service— they do not think, they have not been *taught* to think, of it as spiritual. And here monastic figures who, apparently, found God over the washing up or floor sweeping will be called to mind, but these are not really to the point, since servile tasks were recommended because they left the mind free to contemplate. What we want is a monk who finds God while cooking a meal while one child is clamouring for a drink, another needs a bottom wiped, and a baby throws up over his shoulder.

III

It is not surprising that women philosophers, even when few in number, should have been prominent amongst those who have in recent years criticized that disengaged 'Man of Reason'.[3] Nor is it surprising that many Christian women find themselves little attracted to Augustine and Gregory of Nyssa's spiritual hero who, going outside the feelings his body gives rise to and the vexations of secular life, turns to meet God. Women's lives are much given to attending to particulars; to small and repetitive tasks like the washing of clothes and the wiping of noses that leave no carved stone monuments behind them. Most women in general, if not every woman in particular, have been concerned with the management of ordinary life and the realm of necessity. And most mothers—and indeed attentive fathers—realize that there is something inchoately graced about these dealings.[4] They feel there is something unpalatable about the ancient suggestion that our affection for spouse and children is somehow in competition with a single-minded love of God. That something is unpalatable, of course, does not necessarily make it untrue. Maybe ours is just a spiritually lazy age.

[3] A recent work in this vein is Sara Ruddick's *Maternal Thinking* (1990) in which she attempts, on the basis of what she calls maternal practice, to construct an account of the practical reasoning appropriate to it. Love and attention figure large. The disengaged self, as one might imagine, fares badly as a paradigm of 'attentive parenting'.

[4] I am not making a 'essentialist' case here. It is not that women or 'mothers' are 'born attentive', so to speak, but rather than those engaged in attending to the infant will learn from this. I am using the word 'mothers' for the attenders, pretty much throughout, because, as Ruddick points out, most people who look after children and home in this way are and have been women. This is not to say that they need be women.

Perhaps Gregory is right in thinking the life according to excellence can only be sought via the autonomy he advocates. Mothers of large families would then need to rely on the prayers of these holy individuals bringing benefit to the whole communion of saints.

Certainly, and let me emphasize this now, there is an excellence to the monastic life, and the arguments I have used from Gregory and Augustine are not its only or its best defenses. I am not trying to empty the cloisters, but rather to see what just one of many possible complementary accounts of the spiritual life (there could be many more) for those who do not take this path might be like.

It may be that ours is an unspiritual age, but it may also be that ours is just a different age. In the ancient wisdom of Gregory and Augustine there is a mixture of assumptions of a moral, philosophical and even a scientific nature which we might now want to call into question. For instance, do we think, as Augustine did, that it is 'egregiously silly' to doubt that the life of unchangeable wisdom is preferable to one of change? This no longer seems obvious to us, any more than it is a useful premise for science. We understand ourselves to be creatures of change in a universe that is changing. Cosmology, biology and the social sciences all give accounts of structures, creatures and societies that change. Scientists in general believe that our universe had a beginning and will have an end. Light, hydrogen, carbon, hydrocarbons and primitive life preceded our own human species in this world and made our existence possible. And we might do well to consider wisdom about human beings to be a wisdom about creatures of change.

And again, why should disengagement from the society of the emotional and sensual world be our path to spiritual excellence? Characteristically, as Charles Taylor points out, if we are trying to understand something we aim to be not disengaged but 'fully there', imaginatively present to that which concerns us. It is by this kind of attending that we are characteristically drawn out of ourselves (ecstasis) and come to understand ourselves fully as selves. Central to this are our physical bodies, with all their affective and passible characteristics. Common both to our belief that we are by nature changeable and changing and of necessity creatures of affections is the conviction, unproblematic for most moderns, that we are animals; rational and spiritual animals, perhaps, but for all that in recognizable continuity with other creatures in this universe.

IV

Let us return to the discussion of love and attention with which we began. To be fully human and to be fully moral is to respond to that which demands our response—the other, attended to with love. Morality, religion and mysticism are of a piece.

Let us complement these Platonic themes with an Aristotelian gloss more consonant with our present self-understanding. What we need attend to with love is a changing world full of creatures of change. We ourselves are such creatures.

Let us suppose that affective responses do not, or do not always, mislead and that describing the world as it appears to members of our kind is not inferior to an imagined 'value-neutral' observation of an ideal science but our best handle on the true, the good and the real. Let us suppose that our affections and even our animal responses, properly attended to, are not distractions but guides to what we are and to the love of God.

All life, even plant and protozoic life, is such as to be affected by the world it inhabits. The sunflower turns towards the sun. Attention is rewarded with reality. In his Colour Theory, Goethe makes the following reflection,

> The eye has light to thank for its being. Out of the indifferent animal frame Light has called an organ to be in its own image. And so the eye is built by Light for Light, so that the inner light may encounter the other. (my translation)

Stripped of its teleology this is a point modern biologists make. Not only does seeing 'give us' the world, the world in some real sense gives us seeing. Because of light, organisms have developed photosensitive capacities. (Goethe puts it more beautifully.)

To recapitulate, all life, even protozoic or plant life is such as to be affected by the world it inhabits. Attention is rewarded with reality. But is this a stage to the moral and the spiritual? Not if one thinks that moral and devotional acts are in some stark sense the product of a disengaged reason. Or if one believes that our affections and desires are delusions and snares on our path to the real. Nor yet if one follows what Martha Craven Nussbaum calls Plato's 'double story' with its split between *nous* on the one hand and brute necessity on the other, and the correlative split between human beings and the other animals. In this 'double story' the 'self-moving, purely active, self-sufficient intellect, generator of valuable acts' confronts 'bodily appetites, which are themselves passive and entirely unselective, simply pushed into existence by the world and pushing, in turn, the passive agent' (Nussbaum, 264). A familiar scene but one in which, as it was Aristotle's genius to point out, it is difficult to explain animal motion. But what if, like Nussbaum, we follow Aristotle? Animals, even human animals, according to Aristotle, act on the basis of desires, and the study of animal motion may tell us something important about human ethical aspirations. As Nussbaum says, 'Both humans and other animals, in their rational and non-rational actions, have in common that they stretch forward, so to speak, towards pieces of the world which they then attain or appropriate' (ibid. pp. 275–6). The dog leaps at the piece of meat, both because it desires meat *and* because it sees the object before it as meat. What the Aristotelian account with its focus on desire achieves, Nussbaum points out, is a restoration of the importance of intentionality. It enables us to focus on the intentionality of animal movement, both its 'object-directedness' and 'its reponsiveness not to the world *simpliciter* but to the animal's view of it' (ibid. p. 270). This is not intended by Aristotle to rival the account of the deliberative but to provide 'an attractive account of the natural animal basis for the development of moral character' (ibid. p. 286). Rather than denigrating the animal appetites we should acknowledge that, 'It is our nature to be animal, the sort of animal that is rational. If we do not give a debased account of the animal or a puffed-up account of the rational, we will be in a position to see how well suited the one is to contribute to the flourishing of the other' (ibid. p. 287).

Returning to the spiritual life, once allow our physical natures into the picture as a good, or at least as a necessity, and the vexations of ordinary daily life may appear in a different light. Nothing convicts one more graphically of the implausibility of a sharp distinction between our rational and deliberate capacities, on the one hand, and the bodily appetites and responses on the other, than the experience of pregnancy and attending to an infant. Although it is not everyone's experience to be a mother, it is everyone's experience to have been an embryo and a baby, and it would be surprising if this did not have something to tell us about what it is to be a person. During pregnancy a series of changes take place to the mother's body which make it hospitable to the growing fetus and future child. These range from a suppression of the immune system (so that the fetus will not be rejected) to an increase in the flow of blood and preparation for lactation. None of these changes are voluntary. They are called into being by the presence of the embryo. But consider lactation; in the days following childbirth milk is produced involuntarily in response to the baby's cry. But the mother may be deceived, especially if she is sharing a ward with other mothers and new babies. Imagine this scene where the mother is deceived: the cry is heard, the milk gushes forth but examination reveals that it is not her baby who has cried. The milk stops. Or the reverse: cries are heard. The source is believed to be someone's elses's baby. No milk. A mistake is realized. The milk flows. The important thing to notice is that, in this example, the response of lactation is *both involuntary and rational*, dependent as it is on the mother's beliefs. And this rational component of the maternal response is not discontinuous with the other preparations her body has made. Whereas before birth the mother's body unreflectively attends to the needs of the embryo, after the birth the brain joins the other organs (kidneys, guts, lungs) in attending to the new other. Or better, the whole active being of the mother, in all her instinctual and reflective capacities, is brought to bear on the needs of the baby. Just as at this early stage simple beliefs, such as the belief that it is my baby that is crying, affect simple attentive response to the newborn, so beliefs will become more complex and result in more complex actions—belief that the baby is cold, that babies should get fresh air, that toddlers should be kept from fires, that small children should be courteous to their grandparents, say their prayers and so on. The child is introduced to a world of symbols, stories, goals and practices. By such means parents, even fairly mediocre parents, help babies to become 'selves'.

The process of attending to the child's needs on the basis of parental beliefs is continuous with the simple, involuntary response by which the mother produces milk when she believes her baby is crying. What we do is the result of what we believe about the child, the world, and about what it is good to have and to be. Our affections, though in Gregory's words 'low and animal', are continuous with our highest beliefs and values.

The parents—or those who attend in love—undergo changes as well. The biological reciprocity between mother and child in early infancy is continued in innumerable small acts of watchfulness, many almost as involuntary as lactation. For instance the scanning, native to parents of toddlers, of any new surrounding for steep steps, sharp, breakable or swallowable objects. Parents do not always think much about this, they simply do it as a few years further along in the child's life they will not. Other acts of attentiveness require the disciplined and conscious exercise of what Sara Ruddick calls 'humility'. To attend to the child properly is also to employ the proper passivity of 'letting the other be'. Ideally, as Ruddick says, 'Acts of attention strengthen a love that does not clutch at or cling to the beloved but lets her grow' (Ruddick, 1990, p. 122). By such means parents, at least once in a while, may be 'unselved', just as in Iris Murdoch's example we may be 'unselfed' by the beauty of the kestrel's flight.[5]

The child is, *par excellence*, the individual 'thought of as knowable by love' (Murdoch, 1970, p. 40). Attending to them is a work of imagination and moral effort by which parents try to 'see more' in

---

[5] It should be noted that this 'unselving' is not the destructive abnegation of self which, as Valerie Saiving Goldstein and others have pointed out, is, for many women, a kind of sin rather than a sort of sanctity. A remark of Julia Kristeva's comes to mind, 'The arrival of a child is, I believe, the first and often the only opportunity a woman has to experience the Other in its radical separation from herself, that is, as an object of love'.

Murdoch's sense, or to be 'more fully there', in Taylor's. This is not the moral and spiritual life of the needle-thin, disengaged, 'punctual' self. Rather, as Murdoch says, 'The task of attention goes on all the time and at apparently empty and everyday moments we are 'looking', making those little peering efforts of the imagination which have such important cumulative results' (ibid. p. 43). The object of attention is not a changeless truth so much as a moving target. Children are creatures of change and chance, and an attentive gaze on the real in their case is a gaze on a changing reality.

Despite a certain advocacy for the changeless amongst spiritual writers, the points I have made above are not alien to Christian theology. God is after all portrayed in the Bible as creating a universe that endures for a time but will end. God is represented as attending to a chosen people involved, at God's behest, in seemingly ceaseless change. They are called out, established, exiled, freed. God is represented in the Prophetic writings as chivvying them along and unsettling their complacent accommodations. We seem far from Augustine's tale of the traveller who turns his back on what is material and temporary in order to seek that which is spiritual and eternal, but maybe not. Augustine elsewhere tells another story of a journey, his own journey as recounted in the *Confessions*. We know that Augustine as a young man was appalled by the crudeness of Christianity, his mother's religion, and indeed why should a cultivated man of his place and time have found the stories of an unimportant provincial people like the Jews edifying? But this seems to change once he believes that God became a man and had a human history. The story of the Jews then becomes not otiose and irrelevant but the locus of divine self-disclosure. The history of the Jews, and all human history is, so to speak, 'baptised'. All human history and each human history becomes the place where God meets women and men. Augustine can write his own history of divine encounter, the *Confessions*. For Augustine God's attentiveness does not derogate from God's qualities classically conceived. This is a philosophical leitmotif of the *Confessions*. It is because God is eternal that God is present to all and every time in Augustine's life. God need not be a creature of change to be attentive to changing creatures. God need not be a particular 'thing' to attend to particulars. Indeed for Augustine and for the mainstream of classical philosophical theology, God attends to everything, in particular.

Theologians must reach for analogies. Perhaps the gaze of God is like the gaze of the artist on the completed painting. Each mark of pigment is discrete, yet this green would not be present in its particular greenness were it not for this blue laid down next to it. Each mark has been laid down, one at a time, yet we apprehend the completed work in a single vision. The painting is thus a condensed temporality. We gaze on it as on a complete and consummated whole, bearing all the marks of its making. People might be like this under the attentive gaze of love. Perhaps creation is, too.

References

Augustine. 1988. *St. Augustine's Christian Doctrine*, trans. Rev. Prof. J. F. Shaw. The Nicene and Post-Nicene Fathers, First Series, Vol. II. (Grand Rapids: Eerdmanns).

Murdoch, Iris. 1970. *The Sovereignty of Good* (London: Routledge & Kegan Paul).

Nussbaum, Martha. 1986. *The Fragility of Goodness* (Cambridge: Cambridge University Press).

Gregory of Nyssa. 1979. 'On Virginity', trans. Moore and Wilson. The Nicene and Post-Nicene Fathers, Second Series, Vol. V. (Grand Rapids: Eerdmanns).

O'Donovan, O. 1980. *The Problem of Self-Love in Augustine* (New Haven and London: Yale University Press).

Ruddick, Sara. 1990. *Maternal Thinking: Towards a Politics of Peace* (London: The Women's Press).

Taylor, Charles, 1989. *Sources of the Self: The Making of Modern Identity* (Cambridge: Cambridge University Press).

Descartes' Debt to Augustine

STEPHEN R. L. CLARK

I. More than Cartesian Scepticism

Jonathan Edwards identified the central act of faith as 'the cordial consent of beings to Being in general', which is to say to God (see Holbrook, 1973, pp. 102ff). That equation, of Being, Truth and God, is rarely taken seriously in analytical circles. My argument will be that this is to neglect the real context of a great deal of past philosophy, particularly the very Cartesian arguments from which so many under-graduate courses begin. All too many students issue from such courses immunized against enthusiasm, in the conceit that they have answers to all the old conundrums, which were in any case no more than verbal trickery. 'By uttering the right words but failing to use them *in propria persona*, philosophy induces a kind of soporific amnesia bewitching us into forgetting our God-given task. That task is, of course, to do what Socrates did and to live as he lived' (Burrell, 1972, p. 4). Burrell's words are not wholly fair to academic philosophers, nor to the Lady Philosophy. Plenty of philosophers really mind about the truth, and want to be Socratic in pursuit of it. But the danger is a real one. If all that matters is debunking past philosophers, how does that differ from the repeated refutation of the Chaldaean Oracles or the Prophecies of Nostradamus? A pretty enough pastime for the young, but hardly serious business for adults (as Callicles remarks: Plato, *Gorgias* 484 c 5ff). 'If the history of philosophy is a process of 'salvaging' what you yourself have already thought, then why bother?' (MacDonald Ross, 1985, p. 502).

Descartes is depicted, in those first-year courses, as a failure. Seeking to secure himself against the errors of sense and of scholasticism, he found a sure foundation in the realization, shared by Augustine and a character in one of Plautus' plays,[1] that he was certainly thinking even

---

[1] Augustine, *De Libero Arbitrio* 2.7 (Augustine, 1968 p. 114); Plautus, *Amphitruo* 441–7 (Sosio speaks) (cited by G. Vico, 1988 p. 54). Descartes implies, in a letter to Colvius dated 14 November 1640, that he had not known of Augustine's argument till Colvius mentioned it (Descartes, 1970, pp. 83f), adding that he made a different use of what 'could have occurred to any writer'. Actually Augustine made just the same use of the argument (to establish that we were essentially thinking things) in *The Trinity* (Augustine, 1963, p. 309: Book 10, ch. 10).

when he raised a doubt about that truth, and that therefore someone existed to think that thought. From that uttermost limit of Cartesian scepticism he devised a way of proving God's existence, and hence the general reliability of his senses, when they were treated with a cautious respect. Students are taught that even the *cogito* is unreliable: granted there is a thought (as it might be: 'perhaps there are no substances at all'),[2] but why assume that thoughts must have a thinker? Even if that much is granted, how should we advance from a supposedly clear notion of a perfect being to the actual existence of such a thing? Existence, we learn, is not a predicate, and nothing has a nature such that it must exist *in re*. And even if there were a perfect being, what is to prove that it must speak the truth (unless we already know, or 'clearly conceive', that perfect beings do)?

It would seem to follow that modern analysis leaves us worse off than Descartes. Apparently we are not justified in thinking that we really exist, or that an external or material world exists. None of those theses can be established beyond doubt. We might indeed ('might logically') be the victims of mad scientists or extraterrestrial intelligence. If I *knew* that I was typing this on my PC, I would *know* whatever I saw followed logically from that known fact: as it might be, I *know* that there are no advanced and mischievous intelligences in the universe who love to deceive inhabitants of lesser planets, no Googols (Lehrer, 1970–1; see Clark, 1984, pp. 23f), who are deceiving me in this. But as I plainly do *not* know such a thing (for I know far too little about how often life, intelligence and trickery appear), I do not know even what is most obvious to me. If p, then q: not-q, so not-p. If I knew anything, I would know that no Googols existed to deceive me on that point; I do not, so I know nothing on which they could deceive me. Nor, by parity of argument, do I even 'justifiably believe' any such thing. The only honourable conclusion would seem to be to scepticism of a more than Cartesian kind.

The argument may seem self-refuting: after all, one other thing I do not know is that this argument is valid. But if I do not know that, it does not follow (or I do not see its following) that I have sound reasons for my beliefs. Maybe the hidden premise, that knowledge (and justifiable belief) is transitive, is false: maybe I can know p, but not therefore know any of the things that I can see do logically follow from that p. This is the move preferred by Nozick (1981), but it seems to be as sceptical in its import as the first and swift surrender. If I do not know even what 'follows logically' from a 'known' truth, I hardly know what

---

[2] Otherwise known, perhaps, as the doctrine of *anatta*, the claim that all *dharmas* are empty.

such a known truth means, nor what it means to know it. 'Therefore the wise man will suspend judgement about everything'.[3]

Students are taught to evade, as they suppose, such Pyrrhonism, by remembering that the Cartesian strategy is not for every day. No-one, we insist, can really question the normal, waking world, and draw herself back into the lonely *cogito*. Indeed, we could not even express that *cogito*: self-identification, and the thought that there's a thought expressed, both need the context of a public universe and a public language. To the question, 'am I dreaming?', the answer must be 'no'. No-one seriously doubts her own, and material, reality. But this too is sceptical in its import. If truth is one thing and our words another, there is always a real possibility of error. If we can sometimes truly judge that we were wrong, it must remain an open possibility that we will soon be judging so again—about our present dogmas. The only way of closing off that thought is to insist that we can manage well-enough with what we take to be present agreement, that the limits of our language are the limits of our world. 'Truth', or the only truth that we could mind about, is 'what it is better for us to believe, rather than the accurate representation of reality' (Rorty, 1980, p. 10). That concept 'better', of course, does not mean 'really better', but rather 'what we choose to think is better'. That concept 'we' means only what the speaker and her audience wish it to. Descartes, like other great philosophers, thought it possible that we might discover Truth:[4] present-day critics of his enterprise apparently insist that the Truth we could be wrong about is one we cannot discover. To be 'wrong' in the only important sense is simply to be out of step with our contemporaries (so someone or other says).

The result, which may be very far from any academic's hope, may well be to assure our students that they had better choose to conform, but not of course to assent, to fashionable thought in their own stratum of society. So they bind themselves, as good Pyrrhonian sceptics, to the quadruple compulsion of Nature, desire, custom and the rules of such crafts as they elect to practice, so as not to be wholly inactive.[5] This indolent agreement is so much in accord with fashionable dictats on the dangers of enthusiasm, and smug 'deconstructions' of other people's certainties, as to be well-rewarded. It is also recognizable, at least to those who have sometimes escaped it, as a mild form of depression.

[3] Sextus Empiricus, *Against the Professors* 7.157 (Long & Sedley, 1987, vol. I, p. 255 (41C).

[4] If I capitalize 'Truth' it is not to beg any question about its nature, but only to distinguish it at a glance from the anti-realists' 'truth' (which is only what we choose to say).

[5] Sextus Empiricus, *Outlines of Pyrrhonism* 1.23 (1933, p. 17).

Stephen R. L. Clark

II. Divine Philosophy

So what value can divine philosophy have in this? 'Philosophy teaches us to speak with an appearance of truth on all things, and causes us to be admired by the less learned' (Descartes, 1931, I, p. 86). Descartes was not the first (nor yet the last) to complain that academic philosophizing was betraying a trust. Perhaps we should remember Epictetus' warning, that one who pretends to 'teach philosophy' without the knowledge, virtue and the strength of soul to cope with distressed and corrupted souls, 'and above all the counsel of God advising him to occupy this office'[6] is a vulgarizer of the mysteries, a quack doctor.

> 'The affair is momentous, it is full of mystery, not a chance gift, nor given to all comers. . . . You are opening up a doctor's office although you possess no equipment other than drugs, but when or how these drugs are applied you neither know nor have ever taken the trouble to learn. . . . Why do you play at hazard in matters of the utmost moment? If you find the principles of philosophy entertaining sit down and turn them over in your mind all by yourself, but don't ever call yourself a philosopher.' (Epictetus, 1926, 3.21.17)

When the Lady Philosophy approached Boethius in a death-cell vision it was to turn his mind away from his personal disaster, to awaken in him the spirit that he had professed before, which we still call a 'philosophical' one. Anicius Manlius Severinus Boethius, sometime Master of the Offices, was imprisoned, tortured and clubbed to death in 525 A.D. on suspicion of treason to Theoderic, ruler of the remaining Western Empire. In that death-cell he composed *The Consolation of Philosophy*, of which John Haldane has more to say elsewhere in this volume. The lessons that philosophers ought to rehearse, to write down daily and to put into practice, are the primacy of individual moral choice, the relative unimportance of body, rank and estate, and the knowledge of what is truly their own and what is permitted them (Epictetus, 1926, 1.1.25). 'The Stoics say that every inferior man is insane, since he has ignorance of himself and of his concerns, and this is insanity.'[7] The world of fame and fortune, of everyday concerns, is, so Marcus Aurelius had warned, 'a dream and a delirium',[8] not to be taken seriously except as an occasion for discovering one's own being and the truth. 'Woe to those who turn away from (God's) light and are delighted to cling to their own darkness!'.[9]

---

[6] Epictetus, 1926: *Discourses* 3.21.18.
[7] Stobaeus 2.68, 18 (Long & Sedley, 1987), vol. I, p. 256 (41I).
[8] Marcus Aurelius, *Meditations* 2.17.1.
[9] Augustine, *De Libero Arbitrio* 2.43 (1968, p. 153).

You imagine that fortune's attitude to you has changed; you are wrong. Such was always her way, such is her nature. Instead, all she has done in your case is remain constant to her own inconstancy; she was just the same when she was smiling, when she deluded you with the allurements of her false happiness. (Boethius 1973, *Consolation* 2.1.28).

The philosopher is one who glimpses 'as through a narrow crack' the divine reality (1973, 3.9.8). Till we have done so, and so shaken free of the false glamour and fake terrors of this world, we shall not even be freemen: we are slaves until we are ready to die by torture if it is our job (Epictetus, 1926, 4.1.173). It seems an alarming prospect. Epictetus himself confessed that he and his disciples were, as it were, Jews in word but not in deed: 'not dyed-in-the-wool Jews', very far from applying the principles they teach: 'so although we are unable even to fulfil the profession of man, we take on the additional profession of the philosopher' (Epictetus, 1926, 2.9.21).

Most of us had probably better be content as sophists or as state-kept schoolmen. Epictetus' challenge remains: 'even if you are not yet Socrates, you ought to live as one who wishes to be Socrates' (*Encheiridion* 51.3).

III. Descartes to Augustine

The previous section was intended to make it clear that we do not need to go to distant lands or cultures to discover thinkers strongly inclined to doubt the seriousness or the reality of our everyday existence, and still devoted to the pursuit of truth. Distant lands and cultures should be mentioned more often than they are in undergraduate courses: a brief reading of the texts collected by O'Flaherty (1984) would put an end to much bourgeois philosophistry. But our own Western tradition should have revealed as much to us. 'Really we know nothing, for Truth is in the depths'.[10] The world of our everyday experience, structured as it is by goals and projects of an illusory kind, does not provide us with a sure foundation. We are like children building sand-castles to be washed away,[11] or dreamers. 'For those who've woken up there is one common world; each sleeper's turned aside to a private one'[12]—no less private and delusory because imagined to be shared with others of our greedy, proud and frightened sort.

And what solution can be found? 'We are to stop our ears and convert our vision and our other senses inwards upon the Self'.[13] That is why, so

[10] Democritos (Diels-Krantz, 1952, 68 B 117).
[11] Gregory of Nyssa, PG 44 (cited by Dodds, 1965 p. 11).
[12] Heracleitos (Diels-Krantz, 1952, 22 B 89).
[13] Maximus 11.10b (cited by Dodds, 1965, pp. 92f.

Stephen R. L. Clark

Philo of Alexandria tells us, the High Priest must strip off the soul's tunic of opinion and imagery to enter the Holy of Holies (*Legum Allegoriae* 2.56). This was Descartes' project, and his hope, explicitly, was that he might thereby secure as knowledge what piety already had endorsed. 'For although it is quite enough for us faithful ones to accept by means of faith the fact that the human soul does not perish with the body, and that God exists, it certainly does not seem possible ever to persuade infidels of any religion, indeed, we may almost say, of any moral virtue, unless, to begin with, we prove these two facts by means of natural reason'.[14] Modern commentators neglect our common past because they are conditioned to suppose that 'modern' (post-Enlightenment) philosophers must be estranged from traditional piety, and that 'modern' (positivist) philosophers especially have nothing to say about God or Reality or the Human Soul. When such utterances cannot wholly be ignored in the texts we set our students, we take them to be insincere, and hopeless, efforts to sustain a dying faith. The problem was partly created by Descartes himself. It is a rhetorical trope in very many philosophers to denounce all previous philosophic work. Because students read so little of past history they are too ready to take him at his word, and think that here was where a truly 'modern' sensibility took shape. That Descartes was a theist deserves no more comment than that Newton was an alchemist, or that Copernicus never quite digested what was obvious to Bruno, that the old world was now gone.

What, on my account, did Descartes argue?

(a) The unexamined world we casually inhabit, of lords, priests and commoners, pets and pests and prey and creepy-crawly things, the world of fame and fortune, triumph and adversity, is of the nature of a dream. Nothing that we casually believe, not even the deliverances of contemporary science or contemporary piety, is self-evidently true. Our confidence in them, as long as it rests on unexamined assumptions, is dreamlike in its intensity.

(b) Driving himself downward to his foundations, Descartes concluded that the very act of doubting all things dubitable revealed a Self. The revelation was, so to speak, an existential one, rather than the conclusion of a formal argument.[15] It was not so much that Descartes (or Augustine[16]) identified a certain falsehood in the thought that there

[14] R. Descartes, *Dedication to the Meditations*, 1931, vol. I, p. 133.

[15] The point is made, for Descartes, by Hintikka, 1967; and for Augustine by Mathews, 1972. But it must be acknowledged that Descartes himself *did* appeal to the supposed self-evidence of the claim that 'whatever thinks, exists' (1983, p. 6 (Part I.10)).

[16] See Holscher, 1986, pp. 126ff, for detailed discussion of Augustine's argument for the 'conscious spirituality of the soul'.

78

was no thought, and trusted to the thesis that thoughts required a thinker, but that wondering about existence woke him up.

> Let the mind know itself and not seek itself as if it were absent; let it fix the attention of its will, by which it formerly wandered over many things, upon itself, and think of itself. So it will see that there never was a time when it did not love itself, and never a time when it did not know itself.[17]

That waking self was not self-evidently identical with Descartes: its being lay in thinking (or as a later, Irish, Cartesian said, in perceiving, willing, acting).[18] It must always know itself: 'where could my heart flee from my heart? Where could I flee from my own self?'.[19] If you deny that you have a will, as Augustine observed, you have dropped out of the conversation: 'because I do not have to answer your questions unless you want to know that you are asking. Furthermore, if you have no desire to attain wisdom, there should be no dicussion with you about such matters. Finally, you can be no friend of mine if you do not wish me well'.[20]

(c) the Self revealed is not identical with anything observable, and so is incorporeal. We recognize ourselves as being, not by noticing something that might actually or conceivably be absent, something that is—as it were—'reflected' in the inner eye, but 'really'. 'The known I is simply identical with and equal to the knowing mind itself.'[21] To suppose otherwise—to think that we know only what is 'reflected' as an image in our sensorium would deny us any chance of knowing ourselves to be such a spiritual mirror. Accordingly, I must suppose that I know myself without having, or needing, any 'idea' that is identically me. Since the mind does know itself, and does so in a way that no corporeal entity could, it cannot be a body.

(d) Having discovered the Self, or his self, must the Cartesian be doomed to solipsism? That has been one result, no doubt, and one sometimes blamed on Augustine, 'whose interest was not in the cosmos but in psychological introspection and the question of personal guilt and salvation. He, more than any other Western figure, influenced the Christian West to be individualistic' (Fox, 1988, p. 108). I think the charge unjust. At the least, Cartesians (or Augustinians) do not need to go that way, though perhaps it was true of Descartes, as it was of

---

[17] Augustine, 1963, 10.8,11: see Holscher, 1986 p. 129.
[18] G. Berkeley, 'Philosophical Commentaries' 429 (1948–56 vol. I, p. 53).
[19] *Confessions* 4.7 (Holscher, 1986 p. 132). Aristotle's notion of an intellect that is eternally active (*De Anima* 3.430) may possibly mean much the same.
[20] Augustine *De Libero Arbitrio* 1.24 (1968 p. 95).
[21] Sladecsek 1930, cited by Holscher 1986, p. 291, in 57.

Berkeley and the young Newman, that there were for him 'two and two only absolute and luminously self-evident beings, [him]self and [his] Creator' (Newman, 1967, p. 4). The very act of doubting revealed the Self as sometimes, or as possibly, mistaken. What Descartes realized was that this revealed a truth: namely, that there was indeed a Truth by which his thought was measured. Maybe nothing that I ordinarily suppose is true: so be it, but in that case there is still a Truth, unknown to me, by comparison with which my thought is false. My thought is imperfect, shifting, possibly self-contradictory, finite: but it is all these things because there is a perfect, unchangeable, coherent, infinite reality, *and I already know that it is so*. There could be no doubt, no error, unless there were a Real, nor could we entertain such doubt or recognize the notion of such error unless the image of the Real were stamped within our hearts. As Plato pointed out long before, I must already know what the True is if I am even to notice that my thought might not be true (Plato, *Meno* 80e ff). 'The woman who had lost a coin searched for it by the light of a lantern, but she would never have found it unless she had remembered it',[22] nor known that she had lost it.

(e) of the Truth I only know it cannot be surpassed (as my thought can be), and cannot be denied, or thought not to exist. Even to suggest that it might not exist (that it might not be true that there was such a thing as Truth) is indeed to talk nonsense. '*It is* and that it is impossible for it not to be, is the way of belief, for truth is its companion'.[23] As Edwards put it, time and again:

> It seems strange sometimes to me that there should be Being from all Eternity; and I am ready to say, What need was there that any thing should be? I should then ask myself, Whether it seems strange that there should be either Something, or Nothing? If so, it is not strange that there should BE; for the necessity of there being Something, or Nothing, implies it.[24]

The presence of Being to us is revealed in our discovery that our thought stands under judgment, that our thought is often confused or self-contradictory, but yearns to repossess that which it still remembers. I have an image of Truth (and without it could not even entertain the thought that I am often wrong, that I am not the Truth): it is that eternity of which Boethius spoke, 'the whole, simultaneous and perfect

[22] Augustine *Confessions* 10.18 (1961, p. 224).

[23] Parmenides, as quoted by Raine, 1982, p. 104, after Gerard Manley Hopkins.

[24] Edwards, 1966, p. 51; see also pp. 45f: 'if any man thinks he Can think well Enough how there should be nothing I'll Engage that what he means by nothing is as much something as any thing that ever He thought of in his Life'.

possession of boundless life' (Boethius, 1973, 5.6.9). Because I have that image I both can and must bring my inquiries before it, and accept such clear and distinct ideas as look most like it, always remembering that nothing in the world of my experience or yours is ever quite the Truth (for it might not be true, whereas the Truth itself is always true). 'I had promised to show you, if you recall, that there is something higher than our mind and reason. There you have it—truth itself! Embrace it if you can and enjoy it'.[25] Or as Malebranche put it: 'the truth is uncreated, immutable, immense, eternal and above all things. It is true by itself. It draws its perfection from no other thing. It renders creatures more perfect and all minds naturally seek to know it. Only God can have all these perfections. Therefore, truth is God'.[26]

(f) Descartes therefore believed that there was, that there must be, a way established for us to approach the Truth of which we already know the elements. 'If any other than you were to inspire me', said Augustine to his God, 'I do not believe that my words would be true, for you are the Truth, whereas every man is a liar, and for this reason he who utters falsehood is only uttering what is natural to him'.[27] On the other hand, 'it is really the nature of the soul to live in union with the Divine Ideas, and it depends upon them whenever it pronounces one thing to be better than another'.[28]

IV. Concluding Meditation

What Descartes reconstructed for himself—as would-be believers often must—was the twin discovery that he could deny neither that there was a Truth, nor that he had an image of that Truth. Those who try to deny these truths end by saying nothing: to say that there is no Truth is to say that what they've said cannot be true; to say that the Truth is, or even that it might be, unobtainable, is to say that they could have no epistemic right to say this.[29] So Descartes' conclusion stands.

Critics reply, perhaps, that these are only demands of reason, things that must be supposed true if we are to identify ourselves with the intellectual project of discovering truth. Those who choose to abandon that project will not be caught by Descartes' arguments. To that extent Cartesian epistemology must rest on faith, on a commitment that will

[25] Augustine *De Libero Arbitrio* 2.13.35 (1968, p. 144).
[26] Malebranche, 1980, pp. 234 (3.2.6).
[27] Augustine *Confessions* 13.25 (1961, p. 337).
[28] Augustine *De Libero Arbitrio* 3.13 (1968, p. 177).
[29] The point was also made by the Stoics: see Sextus Empiricus, *Against the Professors* 7.257 (Long & Sedley, 1987, vol. I, p. 246 (40K5)).

endure any number of disappointments and apparent refutations. 'For the Prophet says: "unless you believe, you shall not understand"'.[30] And why then should we hold to that faith, in the power of reason to discover truth by loyalty to the image of the Truth embedded in the human mind? Why not, instead, profess a faith, straightforwardly, in Christian or Islamic or whatever dogma? 'It must be firmly fixed in our memory as the supreme infallible rule that those things that have been revealed to us by God must be believed to be the most certain of all. And that although perhaps some spark of the light of reason might seem to very clearly and evidently suggest to us something else; none the less, trust must be placed solely in divine authority rather than in our own judgement' (Descartes, 1983, p. 36 (Part I.76)). Perhaps the human heart and mind is so corrupt as to have lost, or never to have had, a grip on Truth (or what stands in for Truth). Perhaps such creatures as we are revealed to be by scientific reason never could expect to get a grip on Truth: why should we suppose that creatures evolved with the kind of 'intelligence' they need to get their next meal, or avoid being one, could somehow acquire the knack of discovering their own past history, or the universe's? Revelation is, perhaps, all that we can hope for.

Reasonings like that are what have led self-consciously 'post-modern' thinkers to the kind of pragmatic anti-realism that appeals both to despairing intellectuals and to merely bourgeois anti-intellectuals. According to Lovejoy, 'the primary and most universal faith of man [is] his inexpugnable realism, his twofold belief that he is on the one hand in the midst of realities which are not himself nor mere obsequious shadows of himself, a world which transcends the narrow confines of his own transient being; and on the other hand that he can himself somehow read beyond those confines and bring those external exist- ences within the compass of his own life yet without annulment of their transcendence' (Lovejoy, 1930, p. 14). Rorty, citing this passage, declares (I think absurdly) that Aristotle, Aquinas, Dewey and Austin would (like Rorty himself) reckon this realism artificial and farfetched (Rorty, 1980, p. 52n). There can in the end be no answer to despair except to insist on holding to a hope, and to remark again that anti- realism itself could only be accurate, and properly endorsed, if there was indeed a Truth, and one we could discover. To endorse a universal anti-realism is either to say nothing comprehensible, or to admit its falsehood. That should surely do by way of answer.

To those other anti-rationalists who prefer to keep faith solely with the revealed Word, Augustine has a charitable reply:

Those who exult in divine assistance and who glory in being able to understand and to treat the sacred books without precepts of the kind

[30] Augustine *De Magistro* 37 (1968, p. 50).

which I have undertaken to supply herewith, so that they think these precepts superfluous, should calm themselves for this reason: although they may rightfully rejoice in the great gift God has given them, they should remember that they have learned at least the alphabet from men.[31]

Even those who rely on Revelation must also rest on Reason and Tradition, and need to think through the arguments which identify the possibility of Truth and its Discovery. 'It by no means becomes a philosopher to accept as true something which he has never perceived to be true; and to trust in his senses, that is, the unconsidered judgements of his childhood, more than in mature reason'.[32]

But most students remain puzzled. Suppose I have (as maybe I have not, since 'I am certainly aware of how many figments are born in the human heart, and what is my own heart but a human heart?')[33] established that there must be a real world, Truth unlike my thought in being unchanging, coherent, infinite and necessary. Suppose I have established that we have a notion, an image of that Truth. Why should we identify that Truth with anything we might call 'God'? What does it add to do so, and why do Descartes, and Augustine, think it obvious that what they have established is a perfect God's existence?

Cartesians insist that we have no better model of the Truth than 'God', for we have no better notion of being than as *thinking* being. 'All agree that God is that thing which they place above all other things. And since all those who think of God think of something living, only they can think of Him without absurdity who think of Him as life itself'.[34] Augustine goes on to identify a better life with intelligent life, and then with immutable life: 'for a wise mind which has learned wisdom was not wise before it learned, but Wisdom itself was never foolish nor can be'. It is not that we have equally clear and distinct ideas of the Real as 'personal' and 'impersonal'. Impersonal being is only thought emptied of its significance. Whatever Truth it is that transcends our thought, it can be nothing *less* than thought. It must be that infinite, free spirit in which we live and move and have our being. Edwards again: 'neither Can there be any such thing without consciousness. how is it possible there should something be from all Eternity and there be no consciousness of it. it will appear very Plain to every one that intensely Considers it that Consciousness and being are the same exactly' (Edwards, 1966, p. 48).

---

[31] Augustine, *On Christian Doctrine* (prologue 4) (1958, p. 4).
[32] Descartes, 1983, p. 36 (Part 1.76).
[33] Augustine, 1963, p. 130 (Book 4 ch. 1).
[34] Augustine, *On Christian Doctrine* 1.7f (1958, p. 12).

But precisely because our thought is fleeting, evanescent, submerged in an incessant chatter about this or that, we must also correct many implications of too 'literal' an interpretation of that image. This certainly creates difficulties. 'A mind whose acts and sentiments and ideas are not distinct and successive, one that is wholly simple and totally immutable, is a mind which has no thought, no reason, no will, no sentiment, no love, no hatred; or in a word no mind at all'.[35] Can there be *thought* where there can never be ignorance or uncertainty? Can there be *action* when just wishing makes it so? God, by hypothesis, 'imagines' the cosmos into being, and the being it has is never opaque to Him: it is nothing but what He imagines it as being (which is why Berkeley was correct to see that theism entails the non-existence of a strictly material world).[36] Those who think of God as a spiritual substance (in the sense of 'substance' and of 'spirit' that applies to us) make an error almost as great as those who think of Him as a corporeal one.[37]

Descartes presented his philosophy as a new—or at least his own— way of securing old truths, but there are points on which Cartesian thought does differ from other versions of 'the perennial philosophy'. On the one hand, the effect (though not, I suspect, the intention) of Descartes' argument was to push the evidence of testimony—and experience in general—aside: all that we could be sure we knew was what we could see clearly and distinctly to follow from the central axioms. God would not deceive us if we sought to see exactly what presents itself, but the whole drift of Enlightenment epistemology is to deny us any right to rest content with what we are told about the past. Berkeley, of course, did recognize that there was no substitute for testimony: 'so swift is our passage from the womb to the grave' that no-one could learn much by her own experience.[38] Augustine too had accepted the conclusion:

> In order to contemplate eternal truth in a way that will enable us to enjoy it and cling to it, a path through temporal things, suited to our infirmity, has been marked out for us, namely, that we accept on faith past and future events so far as this suffices for men on their journey towards things eternal. These teachings of faith are so

[35] Cleanthes speaks: Hume, 'Dialogues', 1963, p. 133.

[36] And see Augustine, *Confessions*, 13.38 (1961, p. 346): 'We see the things which you [God] have made, because they exist. But they only exist because you see them'.

[37] Augustine, *Trinity* 1.1 (1963, p. 4). Descartes, 1983, p. 23 (Part 1.51): 'the term "substance" does not apply to God and to those other things "univocally"'.

[38] Berkeley, 1948–56 vol. VII, p. 14 (*Sermon on Immortality*, 1708); see also vol. III, p. 143. See Clark, 1989.

regulated by God's mercy as to give them the greatest authority. . . .
We must readily believe them because they help us very much to
strengthen our hope and to arouse our love.[39]

Although Descartes repeatedly insisted that we could not know what
seemingly possible world God had actually created, and could not
therefore reach the truth by reasonings alone,[40] his actual cosmological
practice was far more dogmatic, and his physical theories as distant
from any real truth as the concluding myth of Plato's *Phaedo*.

On the other hand, Descartes had reasons for saying that God *made*
the laws of logic, and impressed an image of His truth upon our hearts
and minds that yet was not identical with that same Truth. 'They are all
inborn in our minds just as a king would imprint his laws on the hearts
of all his subjects if he had the power to do so' and are such that 'He
could change them as a king changes his laws'.[41] We know *that* the
Truth is (and must be), but *what* it is in itself may indeed be far from
mortal thought.[42] It is necessary to remember this to save ourselves
from the undue literalism that besets so many analytical philosophers.
What we say of Being, of Truth, is not always to be judged as the mere
reproduction in verbal form of a pre-existent Real (though equally, it
sometimes is). The 'truth' of our religious utterances rests in their
evoking and maintaining the 'cordial consent of beings to Being in
general', their being such propositions as 'make proper impressions on
[our] mind producing therein love, hope, gratitude, influencing [our]
life and actions, agreeably to that notion of saving faith which is
required in a Christian'.[43] If we need to recall ourselves to a Truth that
always transcends our thought, and that demands our allegiance as
rational beings, we had best think of it in such terms as do evoke that
attitude, as deserving 'worship'.

What was needed for a purer orthodoxy was the faith that God-for-
us, the image of Truth within our hearts (experienced as laws of non-
contradiction, humanity and reasonable belief), was also God-with-us,
and 'of one substance with the Father'.

[39] Augustine *De Libero Arbitrio* 3.59 (1968, p. 219). See also Descartes,
1983, p. 85 (Part 3.3) on the benefits—and perils—of believing that the world
was made 'for us'.

[40] Descartes, 1983 p. 106 (Part 3.46): 'seeing that [the parts into which
matter is divided] could have been regulated by God in an infinity of diverse
ways, experience alone should teach us which of all these ways He chose'.

[41] Descartes, 1970, p. 11 (to Mersenne, 15 April 1630).

[42] So the claim that the One is 'beyond being' is not as alien to Augustine's
story as (e.g.) Stead (1989) suggests: Edwards', Descartes' and Augustine's
Being, after all, is on the far side of any particular being, and cannot be the
subject of any accidental predicate.

[43] G. Berkeley, 1948–56, vol. III, p. 297.

> Turn where you will, wisdom speaks to you by the imprint it has left on its works, and when you are slipping back into what is outward, it entices you to return within by the beauty of those very forms found in things external. . . . You must come to see that it is not possible to pass judgment, favorable or unfavorable, on things known by the bodily senses unless you have at your disposal a knowledge of certain laws governing beauty to which you refer whatever objects you perceive outwardly.[44]

Without that conviction the One Real remained wholly inscrutable, and we were confined—as later philosophers imagined—to what we were, for natural or supernatural reasons, 'bound' to think, whether or not it was true. The 'image of God', on such terms, would be of an arbitrary tyrant, or a super-Googol, and not such as to awaken love or hope in us. It is significant that Christian orthodoxy and Muslim orthodoxy alike have concluded that the very Word of God is, strictly, uncreated.[45]

The 'inexpugnable faith of man', when properly understood, requires us to have faith in Truth and in the Way of Truth. Those who reject realism are 'of the Devil's party without knowing it', preferring to reign in Hell than serve in Heaven. So how to evade the possibility of Googols, demons, mad neurologists? Why not by confessing, after all, that we *do* know there are no Googols, nothing in heaven or earth, neither principalities nor powers, things present nor things to come, that can separate us from the love of God, revealed in God-with-us? In place of the Enlightenment intelligence, seeking to divorce truth and emotional affect, we should return to the healthier tradition from which Descartes took his start. What's true is what presents itself to those with a lively and a loving faith.

> I am confident that God in His mercy will make me steadfast in all the truths which I regard as certain, but if I am minded otherwise in any point, He Himself will make it known to me, either by His own secret inspirations, or through His own lucid words, or through discussions with my brethren. For this I do pray, and I place this trust and this desire in His hands, who is wholly capable of guarding what He has given and of fulfilling what He has promised.[46]

To understand the world, and to understand Descartes, both require us to remember that realist and Augustinian creed. At the very least, our students might see more clearly that *The Meditations* and *Discourse on Method* are neither sophistical nor easily forgotten.

[44] Augustine *De Libero Arbitrio* 2.41 (1968, p. 151).

[45] See Clark, 1991. pp. 79ff. The point may be the one that Mersenne suggested to Descartes: see Descartes, 1970, p. 14.

[46] Augustine, 1963, p. 9.

References

Augustine. 1958. *On Christian Doctrine*, trans. D. W. Robertson. (Indianapolis: Bobbs-Merrill).

Augustine. 1961. *Confessions*, trans R. S. Pine-Coffin. (Harmondsworth: Penguin).

Augustine. 1963. *The Trinity*, trans. S. McKenna. (Washington: Catholic University of America Press).

Augustine. 1968. *The Teacher, The Free Choice of the Will & Grace and Free Will*, trans. R. P. Russell. (Washington: Catholic University of America Press).

Berkeley, G. 1948–56. *Collected Works*, A. A. Luce and T. E. Jessop (eds.). (Edinburgh: Thomas Nelson).

Boethius. 1973. *Tractates & Consolation of Philosophy*, trans. H. F. Stewart, E. K. Rand and S. J. Tester. (London: Heinemann. Loeb Classical Library).

Burrell, D. 1972. *Analogy and Philosophical Language* (New Haven: Yale University Press).

Clark, S. R. L. 1984. *From Athens to Jerusalem* (Oxford: Clarendon Press).

Clark, S. R. L. 1989. 'Soft as the Rustle of a Reed from Cloyne (Berkeley)', in *Philosophers of the Enlightenment*, P. Gilmour (ed.), pp. 47–62. (Edinburgh: Edinburgh University Press).

Clark, S. R. L. 1991. *God's World and the Great Awakening*. (Oxford: Clarendon Press).

Descartes, R. 1931. *Philosophical Works*, E. S. Haldane and G. R. T. Ross (eds.). (Cambridge: Cambridge University Press).

Descartes, R. 1970. *Philosophical Letters*, A. Kenny (ed.). (Oxford: Oxford University Press).

Descartes, R. 1983. *Principles of Philosophy*, trans. V. R. Miller and R. P. Miller. (Dordrecht & Boston: Reidel).

Diels, and & Krantz, W. (eds.). 1952. *Die Fragments der Vorsokratiker*. (Zurich/Berlin: Weidmannshe Verlagsbuchhandlung).

Dodds, E. R. 1965. *Pagan and Christian in an Age of Anxiety*. (Cambridge: Cambridge University Press).

Epictetus. 1926. *Discourses and Encheiridion*, trans. W. A. Oldfather. Loeb Classical Library. (London: Heinemann).

Edwards, J. 1966. *Basic Writings*, O. E. Winslow. (ed.). (New York: New American Library).

Fox, M. 1988. *The Coming of the Cosmic Christ*. (New York: Harper & Row).

Hintikka, J. 1967. '*Cogito, Ergo Sum:* Inference or Performance?' in *Descartes* W. Doney (ed.), pp. 108–39. London: Macmillan.

Holbrook, C. A. 1973. *The Ethics of Jonathan Edwards*. (Ann Arbor: University of Michigan Press).

Holscher, L. 1986. *The Reality of the Mind*. (London: Routledge & Kegan Paul).

Hume, D. 1963. *Hume on Religion*, R. Wollheim (ed.). (London: Fontana).

Lehrer, K. 1970–1. 'Why not Scepticism?', *Philosophical Forum* 2, 283ff.

Long, A. A. and Sedley, D. N. (eds.). 1987. *The Hellenistic Philosophers*. (Cambridge: Cambridge University Press).

Lovejoy, A. O. 1930. *The Revolt against Dualism*. (La Salle, Illinois).

Macdonald Ross, G. 1985. 'Angels', *Philosophy* **60,** pp. 495–512.

Malebranche, N. 1980. *The Search after Truth*, trans. T. M. Lennon and P. J. Olscamp. (Columbus: Ohio State University Press).

Mathews, G. 1972. 'Si fallor sum', in R. Markus (ed.), *Augustine*, pp. 151–67. (New York and London: Doubleday-Macmillan).

Newman, J. J. 1967. *Apologia pro Vita Sua*, M. J. Svaglic (ed.). (Oxford: Clarendon Press).

Nozick, R. 1981. *Philosophical Explanations* (Oxford: Clarendon Press).

Philo of Alexandria. 1929–62. *Collected Works*, trans. F. H. Colson, G. H. Whitaker *et al*. Loeb Classical Library. (London: Heinemann).

O'Flaherty, W. 1984. *Dreams Illusions and Other Realities*. (London & Chicago: University of Chicago Press).

Raine, K. 1982. *The Inner Journey of the Poet*. (London: Allen & Unwin).

Rorty, R. M. 1980. *Philosophy and the Mirror of Nature*. (Oxford: Blackwell).

Sextus Empiricus. 1933 *Works*, trans. R. G. Bury. Loeb Classical Library. (London: Heinemann).

Sladecsek, F. M. 1930. 'Die Selbsterkenntnis als Grundlage der Philosophie nach dem hl.Augustinus', *Scholastik* 5, pp. 329–56.

Stead, C. 1989. 'Augustine's Philosophy of Being', in *The Philosophy in Christianity,* G. Vesey (ed.), pp. 71–84. (Cambridge: Cambridge University Press).

Vico, G. 1988. *On the Ancient Wisdom*, trans. L. H. Palmer. (Ithaca & London: Cornell University Press).

Visions of the Self in Late Medieval Christianity: Some Cross-Disciplinary Reflections

SARAH COAKLEY

In a volume devoted to philosophy, religion and the spiritual life, I would like to focus the later part of my essay on a comparison of two Christian spiritual writings of the fourteenth century, the anonymous *Cloud of Unknowing* in the West (1981), and the *Triads* of Gregory Palamas in the Byzantine East (1983). Their examples, for reasons which I shall explain, seem to me rich with implications for some of our current philosophical and theological aporias on the nature of the self. Let me explain my thesis in skeletal form at the outset, for it is a complex one, and has several facets.

I. Outline of the thesis

First, this comparison is I believe of some interest, historically and theologically, in its own right, for it witnesses to a fascinating divergence between Western and Eastern Christendom at this point, the West driving wedges between faculties in the self, the East arriving at a remarkable new synthetic view of the person. If I am correct, *The Cloud*, on the one hand, is one manifestation (one amongst the range of possibilities) of an emerging sense of *optionality* in the West in this period about what constitutes the ultimate locus of the self; the *perichoretic* co-operation of memory, understanding and will authoritatively found in Augustine, is, in various ways, rent apart disjunctively in the spiritual texts of this time. No less is Aquinas's integrative view of the 'composite of soul and body' under distinct threat, as we shall see, and here the trend towards a new dualism between soul and body is arguably more manifest in popular spiritual manuals in succeeding generations than in official doctrinal formulations.[1]

As spiritual writers in the late medieval period in the West, then, were driving wedges into classical Christian views of the person from

---

[1] See Copleston, 1955, p. 152, who suggests the same point.

their tradition, Gregory Palamas, admittedly in the face of fierce public opposition from his critics, effected an extraordinary new *rapprochement* between two views of the person that had previously in Eastern Christianity stood somewhat apart, despite a few earlier synthesizing forays. The two views were, first, the Platonizing view of the self as immortal mind, represented by the fourth-century Evagrius, and the more biblically oriented spirituality of the 'heart', 'body' and 'feeling', represented by the fifth-century *Macarian Homilies*.[2] In short, while spiritual texts in the West were *dividing* the person, peeling back to one indivisible point of reference, the East, in its public acceptance of Palamas's viewpoint, was restructuring and resynthesizing it.

On any account, then, this fourteenth-century comparison is of some historical interest (though not to my knowledge previously commented upon in quite this way); and I shall dare to make some speculations a little later about the cultural and social backcloth to such a divergence.

The two other parts of my argument apply some lessons of this comparison to our current debates about identity and immortality in philosophy of religion, debates in which I believe a greater sensitivity to the diversity of Christian traditions in spiritual practice would enrich our sense of *options*.[3] The first point here—my thesis 2—concerns our current theological obsession with the supposed nastiness of Cartesianism, and whether or not we can blame Augustine for it. Rather than taking sides here either with those who find the unacceptable seeds of Descartes' individualism in Augustine, or with those, conversely, who seek to exonerate Descartes in relation to Augustine,[4] I wish instead to divert your attention to the long centuries in between. I want to ask whether the debate amongst medieval historians about whether the twelfth century (or the immediately succeeding period) invented the 'individual' is not of some considerable long-term background significance for our understanding of Descartes' 'sources of the self'. Curiously, even S. Lukes's magisterial little book *Individualism* (1973) only glancingly mentions this earlier period (p. 47). Moreover, current philosophical and theological critics of the Enlightenment understanding of the self have taken almost no account of it; nor have they heeded the claim of Colin Morris (1987)—and others—that it was neither the Renaissance, nor the Enlightenment, that invented the 'individual', but rather the rich period 1080–1150, which developed a peculiar 'self-

---

[2] For this account of Palamas, see Meyendorff, 1964.

[3] My strategy here is not unlike that of Rorty, 1980, chs. 1 and 2, at least in exposing some of the presumptions of our current philosophical debate (see especially the relevant p. 44, n. 13).

[4] For the former position, see (e.g.) Gunton, 1985; for the latter, see Stephen R. L. Clark's contribution to this volume.

awareness and self-expression', a freedom from 'excessive attention to convention or authority' (ibid. p. 7). I shall seek to build on this thesis, and highlight a tendency thereafter, in the later medieval period, to carve back to a unitary, solitary, and to a large extent disembodied sense of the self in the spiral manuals of the thirteenth- and fourteenth-century West. Here are traits which I believe represent interesting foreshadowings of at least aspects of the Cartesian vision, despite all the remaining significant differences.[5]

Finally, and shifting to my comparative example of Gregory Palamas in the East, my third argument—thesis 3—is that Palamas's integrative view of the Christian self, his understanding of the transformative possibilities of the embodied self in this life as an earnest of the next, and particularly his subtle view of the relation of 'body', 'soul', 'mind' and 'heart', all represent options worthy of serious consideration in our debates about personal identity and about life after death. Most important here, perhaps, is the lesson which surely should be more obvious than it is from any cursory survey of historical examples, that we should make no quick identification of 'soul' and 'mind' in our debates about the immortality of the soul. This point is of course already taken by sensitive reinterpreters of Aquinas's view of 'soul' and body; but more usually we are asked to make a choice based on smuggled Cartesian assumptions—*either* the immortality of the soul (=the mind) *or* the resurrection of the body (or neither).[6] Palamas represents just one figure here whose use of 'soul' language is more complicated and interesting than these disjunctions will allow.

II. Cartesian 'individualism'

So much for a brief introduction of my main themes. I want now to include another brief preamble, this time on Descartes (and his modern critics and expositors), for it is important to clarify what *aspects* of 'individualism' we seek to highlight, and which may (perhaps) already be found in the earlier period to which I have alluded.

Fergus Kerr's illuminating book *Theology after Wittgenstein* (1986), for instance, opens with a brief, but fairly devastating assault on Descartes, who represents for him the veritable anti-hero of his whole volume. Descartes' self, he says, 'peels off' everything: 'previous

---

[5] My thesis is obviously less specific than that which links Descartes directly to scholastic predecessors (for which see Gilson, 1979); I concentrate here on what seem significant trends in more 'popular' spirituality.

[6] One thinks, e.g., of the range of views represented in Penelhum, 1970; Lewis, 1973; Hick, 1976; and Swinburne, 1986.

beliefs, senses, body . . . the confidence even that the external world really exists' (ibid. p. 4). The accusations build up on this and succeeding pages: this 'peeled back' Cartesian 'I' is 'completely egocentric'; it is 'self-conscious', 'self-reliant', 'all responsible', but also has a capacity to recede towards ultimate emptiness (ibid. pp. 19–20); it is 'self-disembodying', 'solitary', 'detached', 'objective', 'seeking mastery', even perhaps seeking to be God. Charles Taylor, in his treatment of Descartes in *Sources of the Self* (1989), adds to a similar list of points the familiar charge that for Descartes God is merely an *inference* from '[rational] powers that . . . [he can] be quite certain of possessing' already (ibid. p. 157). 'What has happened', Taylor says, 'is . . . that God's existence has become a stage in *my* progress towards science through the methodical ordering of evident insight' (ibid.).

Now there are an awful lot of different things being claimed against Descartes here, which are often, and confusingly, allowed to heap up under the blanket accusation of Cartesian 'individualism': the usurpation of the divine realm, disembodiment, detachment and solitariness, egocentricity, self-consciousness and moral self-reliance; all these are rather different claims. Indeed, after a survey of such recent secondary views, I am struck by the unanalysed vehemence and passion with which Descartes is sometimes loaded with blame, especially by theologians. What is it, I wonder, about our current social and cultural circumstances, that makes Descartes such an easy target of scorn and loathing? Is it a need to purge out of the system something that we recognize all too clearly in ourselves? To this point I shall return briefly at the end of this essay.

More closely at the level of text, however, I would want to suggest that some of these charges listed against Descartes stand up better under close scrutiny than others. I am not in the business of exonerating him from them *tout court*: that would be foolhardy indeed. But all too often Descartes is read only through selected purple passages, especially the 'peeling back' manoeuvres of *Meditation II*; a more rounded reading of his corpus, and specifically a more contextualized reading of the *Meditations*, reveals some surprising divergences from the archetypal figure with whom we have become familiar. For instance—to give just one or two examples—Michael Buckley (in his *At the Origins of Modern Atheism* (1987)) has elegantly shown that we ignore the intellectual context of Descartes' *cogito* at our peril; Descartes did not just arbitrarily and perversely *invent* his view of the self out of the blue: it arose in response to Montaigne's radical scepticism. The questions and options were thus already set up; Descartes reversed Montaigne's logic and moved *from* the radical scepticism served up to him *to* the certitude of his own existence, *from* the experience of his own imperfection *to* the idea of the all-perfect. 'What is never noted,' says Buckley (ibid. p. 69), 'yet is crucial for the interpretation of [Descartes'] project, is that he did not originate this conjunction between skepticism and the existence of God. It came to him already forged.'

Again, and on another point raised by the critics, the relation of soul and body in Descartes is often badly misunderstood. There is indeed a most radical *distinction* between them, the distinction between the material and the immaterial, the divisible and the indivisible. But Descartes devotes immense energies, especially in his later work, to charting their intrinsically close relation, insisting that the soul resides quite locatably in the pineal gland. Earlier, in *Meditation VI*, he had underscored: 'I am not lodged in my body only as a pilot in a vessel, I am very closely united with it.' And he was very aware, very bothered, that his account of how, precisely, the mind and body interacted, was possibly the weakest and most problematic link in his system. ('This', he wrote to the Princess Elizabeth, 'seems to me to be the question people have the most right to ask me in view of my published works.')[7]

Interestingly, too, under close feminist scrutiny of Enlightenment 'individualism', such as provided by Genevieve Lloyd's account in *The Man of Reason* (1984), Descartes fares—justifiably—rather better than his blatantly misogynist successor in the Enlightenment, Immanuel Kant. Descartes' view of the mind is at least not an intrinsically *sexist* form of 'individualism'. It is however the severe and demanding abstraction of Descartes' form of higher reasoning that bodes ill for those left to (complementary) activity in the kitchen, nursery or salon: 'The life I am constrained to lead', complained the Princess Elizabeth in a letter of 10/20 June 1643, 'does not allow me enough free time to acquire a habit of meditation in accordance with your rules.' Yet Descartes showed some awareness of this difficulty, and even saw its value: 'It is by availing oneself of life and ordinary conversations,' he replies, '. . . and studying things that exercise the imagination, that one learns to conceive the union of the soul and body' (letter of 28 June 1643). There is an ordinary, everyday reasoning, then, that rightly attends to the bodily, the imaginative and the passionate. Furthermore, Descartes is not quite the sustained *solitary* individual that he is often taken for: he does *start* from the individual, and he does say that our interests are 'in some way' distinct from others'; 'none the less . . . one could not subsist alone and is, in effect, one of the parts of the earth, and more particularly, of this state, of this society, of this family to which one is joined' (letter of 15 September 1645).[8]

---

[7] These points are well brought out by Kenny, 1968, ch. 10.

[8] Quoted in Thomson, 1983, pp. 13–14, nn. 2–5, who provides interesting feminist commentary on this theme in Descartes.

Wherein then lies the 'individualism' in Descartes that we seek to expose? It could still, I think we should admit, be located at a number of significant points; but the particular point which I seek to trace back to a much earlier period is this: it is the whittling away of the traditional Augustinian plurality in the faculties of the mind, the reduction to a single mental entity, and thereby the starker ranging of soul over against the body in a somewhat problematic relationship. The point is well illustrated by a passage from Descartes' fascinating late work, *The Passions of the Soul* (Part First, Article XLVII):

> . . . there is within us but one soul, and this soul has not in itself any diversity of parts; the same part that is subject to sense impressions is rational, and all the soul's appetites are acts of will. The error which has been committed in making it play a part of various personages, usually in opposition to one another, only proceeds from the fact that we have not properly distinguished its functions from those of the body, to which alone we must attribute every thing which can be observed in us that is opposed to our reason.[9]

What I propose, then, is to turn back now to the fourteenth century and see if we can find anything like *this* already in process of development at this earlier stage.

III. Twelfth-century 'individualism' and succeeding trends in spirituality

The backcloth here, as I have already mentioned, is the debate about the apparent novelty in attitudes to the self that emerges in the twelfth century (or perhaps, as in Caroline Bynum's refinements of the thesis, a little later).[10] What is at stake here, in Colin Morris's now classic account (1987), is a concatenation of developments associated with the dramatic changes in European society from the period 1050 on, which I can here only summarize blandly and briefly. In this period, the cities assumes increasing significance, education became more widely available, and new classes of educated civil servant, lawyer, critical reformer, and ecclesiastical administrator emerged. In the country the aristocracy, too, was 'transformed in its structures' to allow greater new powers for local lordships, now founded in the principle of primogeni-

---

[9] Haldane and Ross, 1931, vol. I, p. 353. It may be that Descartes has Plato closer to mind than Augustine in his objection to 'any diversity of parts' in the 'soul'; but my point about the *simplification* of the soul still stands.

[10] See Bynum's essay, 'Did the Twelfth Century Discover the Individual?', in Bynum, 1984.

ture. In sum, and to quote Morris, 'Twelfth-century society was thus disturbed by the rapid emergence of a whole series of new groups or classes, all of them requiring an idea on which to model themselves . . . They thus created a conflict of values, and faced the individual with choices which in the year A.D. 1000 would have been unimaginable' (ibid. p. 47). The 'individualism' that Morris goes on to chart he defines initially rather inchoately as 'the sense of a clear distinction between my being and that of other people'. But evidence is then presented of various themes that cluster under this definition: a new sense of self-discovery, inwardness, self-expression, self-examination and intentionality, all to some extent manifest in the ethos and character of the Cistercian reform and in the refinement of the penitential system at this time. Also there was a new freedom to contrast, choose between, and even defy, theological authorities and traditions (well expressed in Abelard's *Sic et Non*; it was one of Bernard's most vehement charges against Abelard that he (Abelard) assumed judgmental freedom over against tradition). And then, most importantly for our interests here, there also emerged a new fascination with the Augustinian faculty of will, reworked in terms of the affections and of erotic desire, which, one might argue, found its secular counterpart in the emerging rituals and attitudes of courtly love.

The crucial point for our purposes here, however, is that writers like Bernard of Clairvaux and his fellow Cistercian William of St. Thierry could give a new, indeed intense, value to feeling in their theology, whilst still, at this stage, holding a balance in their notion of the self between the faculties of knowing and feeling. 'Instruction makes people learned, but feeling makes them wise,' wrote Bernard in one of his *Sermons on the Song of Songs* (V. 14), a remark that might be read as already inviting a disjunction between the two faculties;[11] indeed Bernard does not very adequately explain how love and knowledge are related in moments of mystical union in this life. But he does also insist that love *is* a form of knowing: *amor ipse notitia est* (*De diversis* XXIX, 1); and in another of his *Sermons on the Song of Songs* (XLIX. 4) he describes ecstasy as incorporating both intellect and will.

The importance of all this as backcloth to the later, and significantly different, medieval mystical writers such as *The Cloud* is well brought out in an important recent article by Bernard McGinn (1987). McGinn argues thus. Whilst Bernard, William of St. Thierry, and the thirteenth-century Franciscan Bonaventure all held a balance (albeit somewhat differently enunciated) in their theories of mystical union between the newly emphasized feeling or will on the one hand and intellect or knowledge on the other, an important new sense of disjunc-

[11] So Louth, 1976.

tion between the two faculties becomes apparent in the writing of a figure whose influence lies directly behind *The Cloud*. This is Thomas Gallus, who was the last great mystical writer of the Abbey of St Victor in Paris, who become abbot of Vercelli, and died in 1246. Gallus held the view that there are two quite *distinct* ways of relating to God corresponding to this difference between knowing and loving. There is the *theoricus intellectus*, by means of which we know things that are intelligible, and via them God as first cause; and there is the *affectio principalis*, which is, in contrast, the place where mystical union with God occurs. As McGinn expresses the novelty element here: 'Gallus's language about the relation of knowing to loving in the ascent to God stresses a rather sharp separation or cutting off of all intellectual operations . . . [Thus] For Gallus . . . affective union no longer seems as interested in subsuming the lower forms of intellectual activity as it is in kicking them downstairs' (ibid. p. 13). In short, in my terms, the rending apart procedure, as a distinctive manifestation of later medieval Western spirituality, has here begun. There is a new sense of option: will the closest possibility of relation to the divine fall on the side of affectivity (i.e., will), or on the side of intellect?

It is in this context that we must read probably the most celebrated of the remarks of *The Cloud* author (writing about 1370) on the nature of the self before God. He states (in chapter 4):

> Now all rational creatures, angels and men alike, have in them each one individually, one chief working power, which is called a knowing power, and another chief working power which is called a loving power; and of these two powers, God, who is the maker of them, is always incomprehensible to the first, the knowing power. But to the second, he is entirely comprehensible in each individually . . . (1981, p. 123)

In other words, central to *The Cloud*'s psychology of contemplation is precisely the driving of wedges into the self set up by Gallus: the choice of will over intellect as the faculty of divine interchange. For what 'contemplation' is in *The Cloud* is a 'naked' inclination of the will, a 'peeling back' to this *one* faculty with which the 'sharp darts of longing love' are launched at the darkness of the 'Cloud of Unknowing'.

It is, however, distinctly misleading to regard *The Cloud* as straightforwardly anti-intellectual. When, much later in his treatise (chapter 62 ff.), *The Cloud* author comes to spell out a more technical version of his view of the person, it is a broadly Augustinian package of distinct faculties with which we are presented, ranged under the inclusive rubric of 'mind', though it is a package modified by another Victorine, Richard of St Victor, which *The Cloud* takes over. The 'mind' here is said to 'contain' two principal powers—reason and will, and two secondary ones—imagination and sensuality. Reason (chapter 64) is the power of discrimination, which, since the Fall, is severely perverted. Will, on the other hand (and here we come to the more disjunct tones), is the faculty with which 'we love God, we desire God, and finally come to rest in God with full liking and full consent' (ibid. p. 245). Imagination and sensuality are lower animal faculties, working with the bodily senses.

What we have, then, in *The Cloud*, is a continuing polite bowing to the authority of traditional Augustinian sub-faculties of the mind (later to be rejected altogether by Descartes), but at the same time an entirely new sense of freedom of choice in stripping back to *one*, key mental faculty of divine apprehension, in this case the will. But other mystical authors of the period indicate how free, even bewilderingly free, was the sense of *choice* about the construction of the self at this period of intense mystical flowering. The 'peeling back' could be effected on either side of the will/intellect divide, or even beyond both. Eckhart, for instance, in some moods, chooses to assert intellect over will as the faculty capable of receiving the beatific vision; at other times, as in his treatise *On Detachment*, he speaks of an ascent beyond both knowing and loving (see McGinn, 1987, p. 17). The unifying and simplifying drive to one ultimate locus in the soul, however, is a characteristic trait; and it is no coincidence, I think, though a fascinating detail, that *The Cloud*'s seventeenth-century Benedictine interpreter Augustine Baker later chooses to describe this drive as a process of 'pure and total abstraction',[12] language not used by *The Cloud* author himself, but with distinctly Cartesian overtones.

Two other characteristics of *The Cloud* are also worth mentioning briefly for our purposes. The one is the distinctly problematic, though by no means totally divorced, relationship of soul and body. Fundamentally, in *The Cloud*, as in the later Descartes, the self *is* the soul, and not the body: 'Whenever you read the word "yourself" in a spiritual context, this means your soul and not your body' (chapter 62; ibid. p. 242). Thus 'Every kind of material thing is outside yourself' (ibid.). *The Cloud* does admit that contemplation will—indirectly and secondarily—have a beautifying effect on body as well as soul (chapter 54), and he thinks it is a good idea to look after oneself physically (chapter 41); but he is scornful of those—Richard Rolle is clearly in mind—who look for any kind of delicious bodily effects in contemplation (chapter 52). As an exercise contemplation is a matter for the naked will alone. Secondly, and significantly for our thesis too, *The Cloud* witnesses to another disjunction occurring in this same late medieval period of Western spirituality, and it appears in tandem with

---

[12] McCann, 1924, p. 179.

the split between intellect and will. This is the clear demarcation between 'meditation' and 'contemplation', and the reservation of the latter for an élite group of 'solitaries'. *The Cloud* is a striking witness to this development: the demanding nature of the 'work' of contemplation is to be attempted only by 'specials', he says (chapter 1), who may lead the 'life of the solitary'. This disjunction between meditation and contemplation is a new departure in the West,[13] and not found in the same form in Eastern Christendom; and again, I suggest, it is not without interest in our search for prefigurements of 'individualist' traits in Cartesianism.

But now let me sum up what I have (and have not) claimed in this section in this rather strange and speculative parallelism, before turning to my last point of discussion, on Palamas.

I hope it will be clear that I am not claiming any direct linkage between fourteenth-century mysticism and Descartes: that would be extremely hard, if not impossible, to sustain. In any case, the genres, aims, and contexts we are dealing with are utterly different: mystical treatises on the one hand and intellectual arguments against sceptical detractors (in a distinctly new climate of scientific objectivism) on the other. Moreover, Descartes' 'peeled-back' cognition seems the very inverse of the non-cognitional will of *The Cloud*. My thesis, to repeat, is a more modest, if relatively subtle one, viz. that the new tendencies evidenced in fourteenth-century spiritual treatises (against a wider background of an emerging sense of *choice* over against authoritative tradition) witness to a simplification and reduction of the Augustinian faculties to one, significant locus in the mind or 'soul'. With that goes the loss of the Thomistic integration of soul and body, such that a new and problematic dualism is reasserted; and at the same time the 'contemplative' becomes a professional solitary who recedes, introverts, 'abstracts', from normal practical reasoning for his own particular purposes in relation to God.

Is this not at least fertile ground for later (and in other ways I admit quite novel and different) Western intellectual developments? Certainly nothing quite like this package of assertions about the self emerged in the East, to which I now turn.

IV. The case of Gregory Palamas

I come here to my comparison with *The Cloud*'s slightly older near-contemporary (with whom of course he had no contact at all): the monk-bishop Gregory Palamas (1296–1359). My remarks here are

[13] The development is charted in Tugwell, 1984, chs. 9–11.

necessarily brief and schematic, but I hope none the less suggestive in their comparative force.

Long before he became bishop of Thessalonica (in 1347) Gregory Palamas was a monk of Mount Athos, who spent long hours in 'hesychastic' prayer.[14] This was a type of monastic prayer which had evolved over a number of centuries in Eastern Christendom to incorporate the following characteristics: using a particular posture of the body (resting his beard on his chest), and aligning his prayer with regular breathing, the monk constantly repeated the Jesus prayer ('Lord Jesus Christ, Son of God, have mercy on me a sinner', or some variation thereof). The claim of Athonite hesychasm, which was a large part of what Gregory later had to defend in it against detractors, was that this practice could in time result in the unspeakable joy of seeing the divine 'light' itself, as manifest in Christ's embodied transfiguration on Mount Tabor.

Even thus briefly and crudely described, it will be clear how different this form of prayer is from the 'contemplation' described in such Western treatises as *The Cloud*. Rather than being a prayer of naked stripping and simplification of the faculties, of disembodied yearning upwards into darkness, it has all the reverse features, both literally, and in the use of directional metaphor. That is, it involved a 'positive' use of the body, a pressing of the attention 'downwards' into the body, and a repeated christological emphasis, stressing the unity of body and mind. Note also that the Jesus prayer in the East slides across the West's divide between 'meditation' and 'contemplation': the Jesus prayer can start out as a 'meditational' prayer (in the Western sense, that is, reflection on the theological content of the phrases that are repeated), but can then spontaneously turn into what the West would call 'contemplation', the words becoming a sort of mantric background.

Gregory Palamas's sophisticated defence of this prayer arose because of a vehement assault on its theological assumptions by one Barlaam the Calabrian, a Greek significantly educated in the West, and steeped in the 'negative theology' of Pseudo-Dionysius.[15] Barlaam objected to hesychasm on two counts which naturally sprang from his Dionysian commitment. First, there was the hesychasts' claim to see God in 'uncreated light' in this life; and second, there was the positive use of the body in prayer. Palamas's complex response to the first charge (that is, the impossibility of seeing God in light in this life) involved a rather dubiously-coherent distinction between the 'essence' and 'energies' in God. Thus he attempted to marry and synthesize the darkness mysti-

---

[14] For a brief account of this tradition of prayer, see the sections by Kallistos Ware in Jones, Wainwright and Yarnold, 1986, pp. 175–84, 242–55.

[15] An account is given of this debate in Meyendorff, 1964, ch. 3.

cism of Dionysius and the light mysticism of hesychastic claims, the essential unknowability of the divine and yet the radical availability of God in mystical experience. This argument need not detain us here;[16] we are more interested in the other synthesizing move he made in order to justify the hesychasts on the second charge, that of bodily praying. However, we note that on both counts Palamas treats tradition in a significantly different way from that emerging in the West in this period: one cannot, on his view, pick and choose disjunctively. Spiritual practices have to be justified in terms of an existing rich tapestry of different strands of tradition.

On physical prayer, then, as intimated at the beginning of this essay, Palamas wove *together* the intellectualist understanding of the self as immortal mind (*nous*), as found in the spiritual writings of Evagrius, with the more biblically-oriented psychology of the *Macarian Homilies*, where the self resides ultimately in the 'heart', a term used biblically to denote the psycho-somatic unity of the person. Palamas's core argument in the *The Triads* runs thus (and we note his shifts in terminology between 'soul', 'mind', 'heart' and 'body'):

> Our soul is a unique reality yet possessing multiple powers. It uses as an instrument the body, which by nature coexists with it. But as for that power of the soul we call mind, what instruments does that use in its operations? No one has ever supposed that the mind has its seat in the nails or the eyelids, the nostrils or the lips. Everyone is agreed in locating it within us, but there are differences of opinion as to which inner organ serves the mind as primary instrument. Some place the mind in the brain, as in a kind of acropolis; others hold that its vehicle is the very centre of the heart . . . the great Macarius says . . . 'The heart directs the entire organism, and when grace gains possession of the heart, it reigns over all the thoughts and all the members' . . . Can you not see, then, how essential it is that those who have determined to pay attention to themselves in inner quiet should gather together the mind and enclose it in the body, and especially in that 'body' most interior to the body, which we call the heart. (1983, pp. 42–3)

Thus did Palamas justify, *ex post facto*, the so-called 'descent of the mind into the heart' in the long-term practice of hesychastic prayer. To assess the coherence of this psychology of the person and, not least, account for the precise evocations of 'soul', 'mind' and 'heart', and for the distinctly metaphorical character of the talk of the kenotic descent of the mind into the heart, would be a task going well beyond this present paper. None the less, my suggestion is that this would be a

[16] The theological and philosophical problems of this solution are discussed in Williams, 1977.

worthwhile task for philosophy of religion, and one that would hold up for critical scrutiny an integrative notion of 'soul' at least as sophisticated as that of Thomas Aquinas, and somewhat differently nuanced. But for our preliminary purposes here let us simply note some of the salient additional features of this view of the self. Not only is it clearly rooted in a particular individual prayer *practice*, but it is also integrated into a sacramental communality, and justified in terms of an incarnational principle. Thus, even though, in Palamas, the 'mind' is still seen as the most naturally divine element, the image of God in the person (that represents the Evagrian tradition), none the less the mind has to accept that it cannot save itself by its own powers. It 'needs grace and can find it nowhere but in the Body of Christ united to our bodies by Baptism and the Eucharist'.[17] Thus we note that this rationale acts as an immediate check against an individualistic understanding of the prayer practice. In this context, 'soul' (*psyche*) has the generalized evocation of the life-giving element in the whole person. Hence Palamas can sometimes talk of baptism as representing the 'little resurrection of the soul', by which he apparently means an earnest of the full future resurrection of the soul/body (Meyendorff, 1964, pp. 154–5). 'Deification' in Christ starts in this life, and thus the transformation of the person (body and soul) into the divine life is something that may be already manifest in 'light' now.

V. Conclusions

I have said enough here to indicate how starkly surprising are the contrasts between the newly disjunctive and individualizing trends of late medieval Western spirituality on the one hand, and what emerged as the triumphant Palamite synthesis in the Byzantine East on the other. Part of my plea here, as a theologian with interests in philosophy of religion, is that we should spend a good deal more time than we currently do in our debates on personal identity and theories of life after death analysing the rich *complexity* of spiritual traditions and accompanying theories of the self that Christianity contains; and above all, that we should attend to the way that prayer and sacramental practice (and all their implications) hold together in any implicit such theory of the self. Here I find myself thoroughly in tune with Fergus Kerr's sympathetic interpretation of Wittgenstein on the self (1986, pp. 148–9): how we carve up the self—and for that matter reassemble it—is a matter of actual 'use', and is rooted in a whole complex of social arrangements, rituals, assumptions and ideology.

[17] See the exposition in Meyendorff, 1964, p. 154.

And that brings me to my final point, in a speculative essay in which I have seriously taxed the reader's willingness to make leaps and bounds through time and space. I hinted earlier about the significant correlation in the late medieval West between societal and political change on the one hand, and changes in the conception of the self on the other: readjustments in power centres, the emergence of new classes and groups, the first flexing of nationalistic muscles; all these arguably find their counterparts in the new disjunct possibilities in theorizing about the self. But I have nowhere seen it remarked upon by Palamite scholars that a similar speculation could be made about Palamas's synthesizing *tour de force* of the body/mind. He wrote, we recall, as the Turks, the common enemy, were at the doors of the Byzantine Empire; but a dangerous civil war over imperial succession threatened its internal stability. Palamas, significantly, and in parallel with his treatment of the faculties of the self, was active in effecting reconciliation in the name of the unity of the Christian 'body'. He did this by urging the replacement of imperial and patriarchal powers in the *right place* in the body politic, and by using arguments strikingly parallel to his location of the 'mind' and 'heart' in the human individual (Meyendorff, 1964, p. 64). To draw a final lesson from these examples, then: whilst no suggestion of cultural *determinism* is intended, and whilst debates about identity and the nature of the self must of course continue to be conducted with the utmost philosophical rigour, it is none the less as well to be aware of the political and ideological undertow of our discussions in this area. If theologians seem currently almost united in their collective fury against Cartesian 'individualism', we may well ask what political and ideological agendas (as well as what theology) propel them in this direction. For as the anthropologist Mary Douglas is wont to remind us: 'the public idea of the self is part of a cultural commitment . . . Both self and community have to be examined together.'[18] At the very least, I hope this essay will have provided some revealing examples of that principle, and with material that indicates the rich possibilities for further rapprochement amongst philosophy, spirituality and cultural analysis.[19]

References

Buckley, M. J. 1987. *At the Origins of Modern Atheism* (New Haven: Yale University Press).

[18] I quote from an unpublished paper kindly shown me by the author; but much of Douglas's major published work is devoted to this theme.

[19] I am grateful to John Clark, Andrew Louth, Tarjei Park, Jeffrey Richards and Patrick Sherry for conversations that have helped in the preparation of this paper.

Bynum, C. W. 1984. *Jesus as Mother: Studies in the Spirituality of the High Middle Ages* (Berkeley: University of California Press).

The Cloud of Unknowing (ed. with an intro. by J. Walsh). 1981. (London: S.P.C.K).

Copleston, F. C. 1955. *Aquinas* (London: Penguin).

Gilson, E. 1979 (1913). *Index Scolastico-Cartésien* (Paris: J. Vrin).

Gregory Palamas. 1983. *The Trials*, selections trans. N. Gendle. (London: S.P.C.K).

Gunton, C. 1985. *The One, the Three and the Many: An Inaugural Lecture in the Chair of Christian Doctrine* (London: King's College London).

Haldane, E. S. and Ross, G. R. T. (eds.) 1931. *The Philosophical Writings of Descartes*, vol. I. (Cambridge: C.U.P).

Hick, J. 1976. *Death and Eternal Life* (London: Collins).

Jones, C., Wainwright, G., and Yarnold, E. (eds.) 1986. *The Study of Spirituality* (London: S.P.C.K).

Kenny, A. 1968. *Descartes.* (New York: Random House).

Kerr, F. 1986. *Theology After Wittgenstein* (Oxford: Blackwell).

Lewis, H. D. 1973. *The Self and Immortality* (London: Macmillan).

Lloyd, G. 1984. *The Man of Reason: 'Male' and 'Female' in Western Philosophy* (Minneapolis: University of Minnesota Press).

Louth, A. 1976. 'Bernard and Affective Mysticism', in B. Ward (ed.), *The Influence of St. Bernard* (Oxford: S. L. G. Press).

Lukes, S. 1973. *Individualism* (Oxford: Blackwell).

McCann, J. (ed.) 1924. *The Cloud of Unknowing and Other Treatises by an English Mystic of the Fourteenth Century with a Commentary on the Cloud by Father Augustine Baker, O. S. B.* (London: Burnes Oates and Washbourne).

McGinn, B. 1987. 'Love, Knowledge, and Mystical Union in Western Christianity: Twelfth to Sixteenth Centuries', *Church History* 56, pp. 7–24.

Meyendorff, J. 1964. *A Study of Gregory Palamas* (London: Faith Press).

Morris, C. 1987 (1972). *The Discovery of the Individual* (London: S.P.C.K).

Penelhum, T. 1970. *Survival and Disembodied Existence* (London: Routledge and Kegan Paul).

Rorty, R. 1980. *Philosophy and the Mirror of Nature* (Oxford: Blackwell).

Swinburne, R. 1986. *The Evolution of the Soul* (Oxford: Clarendon).

Taylor, C. 1989. *Sources of the Self* (Cambridge: C.U.P).

Thomson, J. 1983. 'Women and the High Priests of Reason', *Radical Philosophy* 34, pp. 10–14.

Tugwell, S. 1984. *Ways of Imperfection* (London: D.L.T).

Williams, R. D. 1977. 'The Philosophical Structures of Palamism', *Eastern Churches Review* 9, pp. 27–44.

Refined and Crass Supernaturalism

T. L. S. SPRIGGE

I. Introduction

In the postscript to *The Varieties of Religious Experience* William James distinguishes two types of belief in the supernatural, conceived as an essential component in religion, crass or piecemeal supernaturalism, on the one hand, and refined supernaturalism on the other.

For refined supernaturalism the supernatural only provides a totalistic explanation for the world as a whole and never explains any particular fact. This was the typical position of the dominant metaphysics of that time, namely the absolute idealism represented in Britain and America by such thinkers as T. H. Green, Bernard Bosanquet, F. H. Bradley and Josiah Royce, for which everything which is is an aspect of an all embracing eternal cosmic mind or spirit. For these thinkers particular facts are always susceptible of a quite naturalistic explanation at the phenomenal level. The state of mind of refined supernaturalists at that time is well conveyed by Mrs Humphrey Ward's novel. *Robert Elsmere*, which is certainly the best novel on a theological theme of which I know, and in which T. H. Green figures as Professor Grey. (Perhaps the second best is the same author's *Helbert of Bannisdale*.) For the refined supernaturalist special divine interventions, such as miracles, are mere superstition having nothing to do with real spirituality, except as symbols suitable for a more primitive stage of human culture than the present. What this comes to, as James sometimes sees it, is simply that the ordinary natural world of daily life is conceived with a kind of emotional glow upon it. More fairly understood, perhaps, it is the view that statements which seem to be about another supernatural world on occasion intervening in this one are to be understood as a mythical way of saying how human life can be transformed by being lived in a particular spirit.

For crass supernaturalism, in contrast, the supernatural is a realm of concrete existence distinct from our ordinary natural world from which occasional influences emanate. Thus there is a genuine difference between events which have a partly supernatural, and those which have a wholly natural cause. One could, indeed, believe in some such realm without conceiving it as religiously significant, but the crass supernaturalist of James's concern does attach religious significance to it. Thus he will believe in at least some of the following: the occurrence of miracles, the efficacy of prayer in a manner not reducible to facts about its natural psychological helpfulness, in mystical experience as a real form of contact with some supernatural being and perhaps in another world to which he will go after death.

As James says, crass supernaturalism was viewed somewhat dimly by the leading thinkers of his time, and those who were supernaturalists at all were refined ones. However, despite its unfashionableness and rather disreputable air, James declares himself on the side of crass supernaturalism both as regards some of the phenomena it posits and in respect of their religious importance.

He also says that, as a pragmatist, he wants the theistic principle to make a real difference at the level of genuine particular fact. What that difference is is disputable, but minimally it is the real efficacy of prayerful communion with a something *more* than our own consciousness, a communion through which our characters may develop in ways otherwise impossible.

Something pretty close to James's opposition between crass supernaturalism and refined supernaturalism is much alive today. It is true that the more conspicuous players on the side corresponding to that of refined supernaturalism are not absolute idealists today and may indeed repudiate the title of supernaturalists altogether. In fact, in contrast to absolute idealists, their strictly factual or ontological beliefs are, more or less explicitly, of a purely naturalistic character, religious discourse, with its apparent reference to the supernatural, being reinterpreted as pertaining to observances, ceremonies, practices and fables, conceived as uniquely valuable guides to morally good and personally fulfilled living. An early statement of this modern sort of refined supernaturalism is that of R. B. Braithwaite in his *The Religion of an Empiricist*. The work of D. Z. Phillips and Don Cupitt are rather less straightforward examples of something of the same sort. This sort of de-mythologizing theology, in lacking the metaphysical underpinnings of absolute idealism, is certainly in important ways different from that of James's opponents. None the less, there are some close affinites in their attempt to rescue religion from belief in any particular matters of historical fact or from any concrete empirical expectations on the part of the religious believer. In fact, their positions on religion are often remarkably close to that of Bernard Bosanquet in particular.[1] So although they may not like to be called supernaturalists in any sense at all (as indeed neither would many of the absolute idealists) they are, in fact, in the tradition of what James called 'refined supernaturalism' in that they interpret tradi-

[1] See, in particular, Bosanquet, 1889.

tional language about God, the after life, and so forth in ways which dissociate it from any notion of a realm which transcends the natural world in which we exist as perishing physical organisms. It might be better, in fact, to speak of refined and crass spiritualism, rather than supernaturalism, if that expression could be dissociated from spiritualism in the vulgar sense, that being just one extreme and heterodox form of crass supernaturalism.

One must, however, avoid dragooning thinkers into simplified categories and assuming that a thinker one has classified either as a crass or as a refined supernaturalist or spiritualist holds and rejects just what other thinkers thus classified do. For example, this division does not correspond to any division on the importance of ceremonial observance to the religious or spiritual life. James had almost no use for this, though it might seem an essential element in the crassly supernaturalistic aspect of most religions. Modern refined supernaturalists, I take it, are rather likely to emphasize the importance of ceremony and sacraments as elements in the way of life they endorse minus the supernatural trimmings, but Bradley, as a refined supernaturalist, seems not to have had much more concern with it than James. But I believe the classification has real merits. And James and Bradley provide very interesting versions of each point of view, useful for anyone considering the relative importance of each type of supernaturalism or spiritualism for true religion as they understand it.

Both James and Bradley had fathers who were, so to speak, official religionists, James's as a kind of freelance Swedenborgian preacher, associated with American transcendentalism, Bradley's as an Evangelical minister. And while neither adhered to the religious beliefs of his father, each looked to philosophy to provide a defence of an essentially religious or spiritual view of the universe against the apparent threat to it from developing scientific knowledge. It is also worth noting that while each was concerned to develop a theory of things which was essentially spiritual, both emphasized that religion was more a matter of practice than of belief.

There is also a certain similarity in their rather ambivalent relation to Christianity. Both of them believed firmly in the protestant tradition as opposed to the Catholic, but for neither does the figure of Jesus seem to have loomed very large nor do they show any great interest in such distinctively Christian dogmas as that of the Trinity. On the face of it their religious conceptions relate to Christianity only as the most familiar form of monotheism.

Despite these resemblances, they differed profoundly as to what is important in religion.

James insisted that belief in spiritual realities could not be arrived at intelligently by *a priori* arguments. Its support must come from the evidence of religious experience together with that will or right, which he somewhat notoriously argued that we have, to believe in what will be most helpful to us in conducting our personal lives where evidence is not conclusive. However, if an essentially religious conception of the world is to be worth much, it must be more than either a mere habit of conceiving the natural world in the glow of certain abstract metaphysical concepts or even of commitment to a certain style of life. Genuine religion, in fact, depends upon *crass* or *piecemeal* supernaturalism, for which the divine is distinct from the mundane and capable of acting upon it as one agency among others. It was in this connection that James developed his concept of a finite God, who was not so absolutely in control of things as to be responsible for all evil, and to stand in no need of assistance in the fight for good, as would be the case with a truly omnipotent God. Such a finite God, whose victory in the world is not guaranteed, but which we can freely participate in making probable, seemed to James both morally and evidentially superior to that of an absolutely omnipotent God. He also thought it nearer to the real beliefs of the ordinary religious believer.

Bradley's position was almost totally the opposite. A spiritual view of the universe was, for him, at least *qua* philosopher, something arrived at on the basis of metaphysical argument. He despised attempts to base it on alleged particular matters of fact. Thus in his first published work he was as scathing upon the topic of miracles as Hume. And while James believed that spiritualistic phenomena (in the vulgar sense of those supposed to occur through mediums) might support a religiously relevant piecemeal supernaturalism, Bradley thought that, even if true, they had nothing of religious value. Moreover, Bradley's Absolute, an all embracing universal consciousness containing all finite things within it, might well be thought the apotheosis of that metaphysical substitute for a living God, which James sought to combat through his idea of a finite one. (What is more Bradley at least tends to treat 'all evil as partial good' in the manner which James deplored as an implication of monistic idealism, though Bradley himself thought orthodox theism worse in that regard.)[2]

There are, indeed, some similarities among the differences. For example, James's commitment to crass or piecemeal supernaturalism does not seem to have gone with any particular interest in the Gospel miracles, the virgin birth, the resurrection and so forth. Likewise they both thought that religious belief was primarily a practical and ethical matter, and that religious creeds were to be evaluated, not for their

---

[2] This emerges particularly in a letter (the relevant part of which is unpublished) to William James dated 21 September 1897. (It is in the William James collection, Houghton Library, Harvard.)

ultimate metaphysical truth but as working ideas stimulating the spiritual and moral life. From this pragmatic point of view, Bradley was even prepared to grant that there was some place for thinking of God as having goals not guaranteed of fulfilment.[3] However, he held that this must be combined, even at the level of ordinary religious working conceptions, with a sense that the final goal was guaranteed, and indeed somehow already eternally reached. For the idea that the supremacy of good in the world was only a possibility genuinely at risk seemed to him incompatible with an adequate religion.[4]

II. Bradley's Views on Religion

(a) *His Views on Gospel Miracles, etc.*

Bradley's negative attitude to the supernatural element in Christianity, in the popular sense of the word, is clear in his very first published work, *The Presuppositions of Critical History* of 1874. In it he discusses the circumstances under which the critical historian may accept historical records as factually correct. Its main thesis is that all belief must be an interpretation of our own experience on the basis of our own preconceptions, and that this means acceptance of the testimony of historical witnesses only to the extent that we can work out what we would ourselves have concluded from their experience. Thus one should take their testimony at face value only to the extent that their outlook is rational by our own lights. Consequently there are strict limits to the extent to which we should accept records of events which clash with our own preconceptions (Bradley, 1969, p. 31).

His argument has some similarity to Hume's treatment of miracles. However, Bradley is careful to avoid the implication that testimony cannot give one grounds for believing in facts of a type one previously thought impossible. What one must not do is accept testimony to such an effect when it comes through minds with whom one cannot identify and presume to have sifted things as one would have done oneself.

Although written in the context of controversies over the 'higher criticism' of his day. Bradley does not say anything directly about the acceptability of the traditional accounts of the historical Jesus (or of Old Testament narratives) though this was presumably his main concern. But it emerges clearly enough that he was quite unsympathetic to forms of Christianity for which particular historical claims are central. In so

[3] A letter to James of 14 May 1909 has some bearings on this. See Perry, Vol. II, pp. 638–40. See also Bradley, 1968.

[4] See, for example, Bradley's letter to James of 14 May 1909, partly reprinted in Perry, Vol. 11, pp. 638–40.

far as Christianity requires belief in such events as the resurrection he was quite alienated from it. Thus the only sort of Christianity in which he had a positive interest was that which offered an interpretation of the significance of human life in whatever particular events it unrolls.

Philosophers tend to divide into those, like Hegel, for whom reality in its most important aspects is essentially historical, and those, like Schopenhauer, for whom the essential facts about all time are much the same. It is not too easy to classify Bradley in either way. On the one hand the historical is the concrete and it is only the concrete which is fully real for Bradley. On the other hand, the unrolling of events in time only matters in so far as an unchanging eternal reality shines through them appreciation of which cannot be tied to knowledge of any particular piece of world history.

(b) *His Views on an After Life*

Another aspect of Bradley's insistence that true religion has little to do with supernaturalism in any popular sense is manifested in his attitude to the belief in life after death, and particularly to the attempts of spiritualists to prove it empirically. His main conclusion on these topics is that whatever the truth may be it is not religiously very important.

Treated as a purely factual question Bradley seems always to have regarded the issue of personal survival as open. He insisted, however, that the truth may not be expressible as a stark 'yes' or 'no'. And the view that the question is not of primary importance for religion, or indeed for metaphysics, seems to have been constant with him.

In an article published in 1885, called 'The Evidences of Spiritualism'[5] he offers a critique of some claims made in *Miracles and Modern Spiritualism* by A. R. Wallace, the evolutionary thinker whose propounding of a doctrine of natural selection hastened Darwin's publication of *The Origin of Species*. Bradley says that the spiritualist must prove three things; first, that his 'phenomena' are real, secondly, that they are not wholly the result of abnormal powers on the part of living human beings, and thirdly that, even if they are the work of spirits from another realm, these spirits are *our dead*. And Bradley insists that, even if we granted that the first two had been demonstrated, the spiritualist would still be in no position to establish the third. For one thing, he has never faced up to the problem of what can show or even constitute personal identity when the continuities of the natural world are broken.

In any case, just what sort of souls are these which manifest themselves in spiritualistic phenomena if there really are such? After some

---

[5] Chapter XXIX of Bradley, 1969.

amusing banter as to whether their physical skills are or are not superior to our own, Bradley continues by arguing that all the evidence is that they are, at least intellectually and morally, vastly inferior to us.

> To the damning evidence of the so-called Spirit-Teachings no answer can be made. It would be unfair to say that the best of them are twaddle, and they perhaps may be compared with our own pulpit-utterances. They are often edifying, and often reasonable, and sometimes silly, and usually dull. Still, to mention them in the same breath with the best human work would be wholly absurd. And it is an inferior race which can produce nothing better. (Bradley, 1969, p. 600)

He concludes that even if spiritualistic phenomena were proved genuine, any religion based upon them would 'conflict with the best aspirations of the soul in a way in which modern materialism does not', inasmuch as it would be concerned with communication with spirits inferior to ourselves which we must hope are not in any sense (and there are considerable problems in knowing what such sense would be) what we become on death. If there is a spirit world it may well be so inferior a world that humanity has risen precisely because it has detached itself from it. All things considered, it is more congruent with genuine religion to believe that after death we continue only in what we have achieved in this life.

There is a brief and eloquent discussion of the issue of an after life in *Appearance and Reality*, 1893.[6] The conclusion is that some form of survival, whether with or without a body, is a bare possibility, but no more. He repeats his rejection of spiritualistic evidence, emphasizing even more strongly than before that it is far from established that there is communication here with spirits of any sort, let alone with our dead. As for the usual metaphysical arguments they are all radically flawed. The best argument would be one which showed that the perfection of the Absolute, which Bradley sees as a metaphysical necessity, requires our personal survival, but no such argument holds water. The universe has needed me in this life, but why should it need my survival? It may be said that, without it, virtue would have no adequate reward and evil no sufficient punishment. Quite so, but all we can be sure of, on the basis of Bradley's doctrine of the Absolute, is that the evil in the world is somehow made good. To assume that it must be made so for the particular person who suffers it, gives a metaphysical centrality to the individual person pertaining to an inferior form of religion which we should move beyond. The best side of a belief in a life after death lies not in the concern we may have at our own cessation, but in the wish we

---

[6] Bradley, 1930, pp. 444–52.

have to see our loved ones again. But this could be a mixed blessing, when we consider the complications of relationship which arise between those who survive after a loved one's death. Still, if it comforts someone to believe that there is a bare epistemic possibility that there is a life after death, so much can certainly be granted him. But it can corrupt both religion and morality to rest their validity on a belief which upon the whole the educated world is losing.

A rather more positive tone towards the possibility of life after death is adopted in some of the later essays published as *Essays on Truth and Reality*.[7] Altogether these essays show more sympathy with the other worldly note in religion than the earlier writings.

He does, indeed, still insist that 'mere personal survival and continuance has, in itself, absolutely nothing to do with true religion. A man can be as irreligious (for anything at least that I know) in a hundred lives as in one' (Bradley, 1968, p. 440). But if a doctrine of personal survival is the one way in which people can formulate the conviction that, at our best, we are identical with something which transcends time, then he is not without sympathy with it.

In 'On our Fear of Death and Desire for Immortality',[8] one of Bradley's most moving pieces, he discusses the lack of precision in the whole question of personal survival. For it is not at all clear what has to exist in a future time for it to be true that I survive. Of all the various possibilities, however, the only one in which I can take anything properly called a personal interest is one in which some future individual looks back upon me as I am now with a special personal interest (Bradley, 1968, p. 45).

Upon the whole Bradley thinks that a special concern with one's own future rests upon an illusion. And the fear of annihilation may arise partly from the confused belief that it is an extreme version of those forms of partial annihilation, for example in senility, which really are objectionable (ibid. p. 456).

But may not lovers wish to survive as lovers for ever? Rather, thinks Bradley, in the most passionate love the question becomes irrelevant.

> Something has been revealed which is beyond time and sports with the order of events. There never was a before, and God has made the whole world for this present. (ibid. p. 457)

Thus in moments of completest fulfilment we live as though in an eternal now, and that moment is, then, indeed one of the eternally

---

[7] See especially Bradley, 1968, Chapter XV, including Supplementary Note B, and Chapter XVI.

[8] Bradley, 1968. Chapter XV Supplementary note B.

present peaks of the cosmic experience in which everything has its place. This theme is taken up again in 'On My Real World'.

'For love and beauty and delight', it is no matter where they have shown themselves, 'there is no death nor change'; and this conclusion is true. These things do not die, since the Paradise in which they bloom is immortal. That Paradise is no special region nor any given particular spot in time and space. It is here, it is everywhere where any finite being is lifted into that higher life which alone is waking reality. (ibid. p. 469).

Such passages in Bradley's later writings (anticipative of the *Four Quartets*) exhibit an increasingly Platonic insistence that the world of phenomenal flux is of value only for the eternal forms which are actualized in it from moment to moment.

In all this Bradley is very much the refined supernaturalist criticizing the crass supernaturalism of which James was the avowed champion (James, 1985, pp. 520–1). There may be a crass supernatural, Bradley allows, but what is spiritually significant is as present in this world as it could be in any other. In all equally there can be the expression of goodness, truth and beauty. Bradley, of course, believes in an absolute cosmic consciousness which lives in us all, but what is important in the cosmic life certainly need involve no supernatural events in the popular sense. Let us now turn to James.

III. James on Religion

For James (for the most part) religion stands or falls with the genuineness of crassly supernatural interventions into the mundane world or at least of the influence upon human consciousness of something belonging to another level of being from the natural. What seemed to him of most importance in this connection was putative mystical experience. But he was also actively interested in psychical research, in particular, in the activities of spiritualistic mediums. He was never absolutely convinced here but remained a sympathetic though critical investigator. However, unlike Bradley, he believed that the establishment of the validity of these phenomena would have religious significance. He did, however, agree with Bradley that it was not life after death which was of prime religious importance. It was rather simply that there should be forces at work in the world whose doings could not be explained in a purely naturalistic manner and that some of these, at least, be powerful influences for good with which we can unite ourselves.

Let us investigate his approach to mystical experience and to spiritualistic phenomena in turn.

T. L. S. Sprigge

(a) *The Argument from Religious Experience*

In the *Varieties of Religious Experience* James, as is well known, distinguishes the once born and the twice born type of religious consciousness. The former experiences a joyous sense of identification with the divine, without any prior passage through the valley of the shadow of death, while the latter only reaches it after passing through a sense of abandonment and desolation. For the first, mystical experiences are simply the peak of their typical way of being in the world, for the second, they come as offering salvation from a complete wreckage of the soul. In both there is the same sense of being somehow in union with a greater consciousness than their personal one. This takes its most humanly valuable form when it is the inspiration for saintliness.

Mystical experiences are essentially ineffable, but one can say, at least, that they carry the sense of being taken over by some larger power from which saving experiences come. And they do not feel to their subject like merely subjective feelings. Rather do they carry the irresistible conviction that he is somehow continuous with a larger whole with which they bring him into a fuller than normal relationship. Thus he feels grasped and held by a superior power. And they bring an intense sense that reality is one and good, and that the individual is simply an element in this divine unity.

Thus mystical experiences come with an apparently noetic or cognitive quality, presenting themselves as informative about reality (James, 1985, pp. 380–81). And the information they seem to bring is typically of a monistic, pantheistic and optimistic character, though this may be downplayed by those who interpret it in terms of a religious orthodoxy which this would threaten.

Of course, the optimism of the twice born is not of the easy type of the once born. Rather is it the conviction that the apparent badness of the world, and of oneself, is not the last word. Still, in their different ways, mystical experiences of both types seem to carry the message that the totality of things is supremely good.

That, at least, is the message they seem to bring with such force to their subject. And such a monistic interpretation is certainly the initially most natural one of what is going on. But we are not, thinks James, bound to interpret the deliverance of these experiences in quite such a strongly monistic way even if we acknowledge them as genuinely noetic. For the core content of the mystical experience is the sense of being part of, or at least continuous with, a divine life. This does not require that the divine reality be the only reality, and is compatible with its having a resistant other, which we may join with it in fighting. Either way, if mystical experience is not a delusion, there really is a larger world of consciousness than that with which the ordinary facts of nature acquaint us, and the best solution to our woes lies in entering into felt unity with it.

But mystical experiences are of various types which may call for different explanations. All point to a reality beyond that open to our senses or ordinary thought. But this reality may be complex and various, and different sorts of mysticism may bring contact with different elements of this larger world. Some mysticisms, indeed, seem to put people in touch with the diabolic rather than the divine. But all are either delusive or forms of contact with a larger and more spiritual world than that of ordinary life.

> It is evident that from the point of view of their psychological mechanism, the classic mysticism and these lower mysticisms spring from the same mental level, from that great subliminal or transmarginal region of which science is beginning to admit the existence, but of which so little is really known. (James, 1985, p. 426)

Still, most mystical experiences are of continuity with something appropriately called divine. But do they establish satisfactorily, in a way which should bring conviction to all reasonable persons, that there really is such a reality? James summarizes his answer thus:

> (1) Mystical states, when well developed, usually are, and have the right to be, absolutely authoritative over the individuals to whom they come.
> (2) No authority emanates from them which should make it a duty for those who stand outside of them to accept their revelations uncritically.
> (3) They break down the authority of the non-mystical or rationalistic consciousness, based upon the understanding and the senses alone. They show it to be only one kind of consciousness. They open out the possibility of other orders of truth, in which so far as anything in us vitally responds to them we may freely continue to have faith. (James, 1985, pp. 422–3)

So we can sum up the purely evidential aspect of the matter, as James sees it, thus. Mystical experiences inevitably seem to their subject to establish that there is a divine reality of some sort with which he has had contact. But it does not follow that this is really established thereby. The most it can reasonably force the sceptic to acknowledge is that he has no right to complete confidence in the adequacy of a purely naturalistic account of the world and that the religious person has as much of a right to his vision of the universe as the materialist has of his. The existence of mystical experience should check any 'premature closing of our accounts with reality'.

T. L. S. Sprigge

(b) *James on Psychical Research*

James's reputation suffered in his life through his commitment to psychical research, and it may do so with some people today (Bjork, 1983). But if he seems credulous to some, he certainly never showed the extreme gullibility of such a contemporary as Conan Doyle (Jones, 1989), and, in fact, he shows an admirable combination of open mindedness and caution.

The best summary of his attitude is contained in the 'Impressions of a Psychical Researcher' of 1909. The article starts with a reference to Henry Sidgwick's remarking that, if he had realized from the start that, after twenty years' engagement in psychical research, he would have still been in his same original state of indecision as to the validity and significance of its phenomena, he would have been amazed. James says that his state is similar after his own twenty-five years of investigation, direct and through the literature. Still, he cannot take seriously the suggestion that the whole mass of mediumistic phenomena is sheer fraud. Sometimes, indeed, he wonders half seriously whether the difficulty in determining which, if any, phenomena are genuine, and why they tend to be so absurd in character, may be due to the Creator's wish for 'this department of nature to remain baffling' (James, 1924, p. 175).

As for his final conclusion, it is that, amidst much humbug, at least some of the performances of mediums rest on genuinely supernormal knowledge. As to how this should be interpreted James remains very open-minded. He certainly does not think that they prove personal survival. A certain will to personate is a part of normal human psychology and mediums may unconsciously personate dead persons about whom they gain supernormal knowledge from a source which cannot properly be said to be the dead person. But this is speculation. James remains simply a psychical researcher, waiting for more facts before concluding.

Yet he does draw a vague general conclusion of considerable interest from the point of view of crass supernaturalism.

> Out of my experience such as it is (and it is limited enough) one fixed conclusion dogmatically emerges, and that is this, that we with our lives are like islands in the sea, or like trees in the forest. The maple and the pine may whisper to each other with their leaves, and Conanicut and Newport hear each others' foghorns. But the trees also commingle their roots in the darkness underground, and the islands also hang together through the ocean's bottom. Just so there is a continuum of cosmic consciousness, against which our individuality builds but accidental fences, and into which our several minds plunge as into a mother-sea or reservoir. Our 'normal' consciousness

is circumscribed for adaptation to our external earthly environment, but the fence is weak in spots, and fitful influences from beyond leak in, showing otherwise unverifiable common connection. Not only psychic research, but metaphysical philosophy and speculative biology are led in their own ways to look with favour on some such 'panpsychic' view of the universe as this. Assuming this common reservoir of consciousness to exist, this bank upon which we all draw, and in which so many of earth's memories must in some way be stored, or mediums would not get at them as they do, the question is, What is its own structure? What is its inner topography? . . . What are the conditions of individuation or insulation in this mother-sea? To what tracts, to what active systems functioning separately in it, do personalities respond? Are individual 'spirits' constituted there? How numerous, and of how many hierarchic orders may these then be? How permanent? And how confluent with one another may they become?

What again are the relations between the cosmic consciousness and matter? Are there subtler forms of matter which upon occasion may enter into functional connection with the individuations in the psychic sea, and then only show themselves?—so that our ordinary human experience, on its material as well as on its mental side, would appear to be only an extract from the later psycho-physical world?

Vast, indeed, and difficult is the inquirer's prospect here, and the most significant data for his purpose will probably be just these dingy little mediumistic facts which the Huxleyan minds of our time find so unworthy of their attention. But when was not the science of the future stirred to its conquering activities by the little rebellious exceptions to the science of the present? . . . (James, 1924, pp. 204–5)

(c) *The Will to Believe and the Pragmatic Conception of Truth*

In treating James as a crass supernaturalist for whom a religious conception of the world requires belief in crassly supernatural events occurring as a matter of objective fact, it may seem that I am altogether disregarding his pragmatism and his doctrine of the will to believe. For surely these imply that a religious view of the world is to be adopted if it is emotionally helpful and morally inspiring, and can be called true, if it is so, without overmuch concern for what fact-grubbers may call the objective truth of the matter.

It is as well to distinguish the will to believe doctrine from pragmatism, which was a later thesis. Let us take pragmatism first[9]. There are

[9] The main statements of this are James, 1937 and James, 1975.

certainly strands in pragmatism which could support a refined supernaturalist account. For pragmatism claims that true ideas are not
necessarily those which provide us with a symbolic transcript of a pre-
existent reality but those which enable us to live effectively in the
world. That suggests that if religious narratives, including the supernatural events they typically include, are such that the entertaining of
them encourages us to live in a more satisfactory manner, then they
have the kind of truth suitable for religion, even if not for scientific
historical claims, which are designed to provide a different sort of
satisfaction.

However, though pragmatism can be developed in that direction it is
not James's. For James's pragmatism is basically a theory about how
symbols come to have meaning, and says that they mean that with
which, if it exists, they enable us to cope successfully. It does not deny,
but rather implies, that ideas are true only if there is a fact to which they
correspond, only it defines that correspondence in terms of ability to
adapt us towards the fact. This leaves it quite open whether religious
ideas are to be interpreted in a refined supernaturalistic or a crass
supernaturalistic manner. If the facts which must exist if religious ideas
are to work are facts about the character of a purely natural world, facts,
for example, about human psychology and society, then their
apparently crassly supernaturalistic sense is not their real meaning. If,
on the other hand, what they serve to adapt us to are emanations into
the natural world from some distinct supernatural source, then their
true meaning is crassly supernatural. And upon the whole, though, I
admit, with some exceptions, that is how James interprets them.

As for the will to believe doctrine this tells us that where cognitive
evidence and demonstration leave a matter undecided we have a right to
take a view of how the facts are on the basis of our emotional needs. But
it does not deny that there is a fact of the matter in virtue of which we are
objectively right or wrong.[10]

Mystical experience makes religious belief a serious cognitive option
but does not prove its claims. Likewise the success of purely naturalistic
accounts of so much that was once explained by divine causation
favours, without establishing, a purely naturalistic world view. That
leaves many of us, so James thought, in precisely that situation where
the will or right to believe has place and we have the right to commit
ourselves to whichever one of a pair or set of alternative beliefs, which
will best serve our passional nature, that is, our emotional and volitional
needs, whenever there are no adequate empirical or rational grounds,
and the option between them is momentous, forced, and live.

---

[10] See James, 1923. Later he wished he had called it 'The Right to Believe'.

An option is *momentous* when it matters greatly to us in which alternative we put our faith, *forced* when our style of life will inevitably express one of the alternatives, however, much we decline any express affirmation, and *live* when each seriously solicits our assent. These are, in fact, the circumstances in which James thought his contemporaries typically found themselves when they hesitated between a religious and a naturalistic view of the world. So he recommended them to follow him in making the choice on the basis of the promise which each view holds of being helpful morally, volitionally and emotionally. (It is in this spirit that *The Varieties of Religious Experience* investigates, and endorses, the claims of saintliness, and other human products of religion, to be the greatest, human good.)

What then is the nature of the religious options which face our passional nature and between which we may choose, so far as each passes muster evidentially according to how we value the particular emotional and volitional goods, such as peace of mind and moral vigour, which each offers? Actually James saw the options as fourfold rather than twofold: (1) orthodox theism; (2) materialism or naturalism; (3) pantheism; (4) the doctrine of a finite God.[11]

1. For James the first option, orthodox theism, was decreasingly serious.[12] As advocated by scholastic philosophers it became a logomachy with little hold on living reality or on the hearts of men. In more recent times, indeed, orthodox thinkers relied more on empirical considerations than on the *a priori* arguments of the scholastics, appealing primarily to the argument from design (James, 1985, p. 438). This was certainly an improvement, as James saw it, for it makes the existence of God a hypothesis for the explanation of undoubted empirical facts, such, in particular, as the adaptedness of organisms to their environment. Unfortunately for orthodoxy, however, it is no longer the best hypothesis on offer, since the phenomena are better explained on Darwinist principles.

Traditional theism also suffers from its incapacity to deal adequately with the problem of evil, a problem which seems more serious to the more sensitive modern mind.

For these reasons orthodox theism is no longer a live option for many of us.

2. Naturalism or materialism,[13] James recognized, was certainly a living option then as never before.

---

[11] See, in particular, James, 1985, chapter XVIII and James, 1909, chapter one.

[12] See James, 1985, pp. 436–442 and PU p. 29.

[13] By 'materialism', at that time, was usually meant epiphenomenalism; the absurdities of the modern 'identity' theory were hardly dreamt of.

As James saw it, such a view of the world was favoured both by Darwinism and by increasing knowledge of the dependence of consciousness on brain process. Moreover, our increasing sensitivity to the horror of so much human suffering makes it more difficult for us to see the world as having been created by a perfect God. Altogether, then, there is much to dispose modern man to this view of things.

However, James thought such a view of the world by no means proven. It is at most a legitimate hypothesis covering many of the observed facts very well. But while religious experience, and perhaps some of the phenomena investigated by psychical researchers, do not conclusively falsify it, any one of the other hypotheses, each of which is in its own way religious, seems *prima facie* in a better position to make sense of them.

3. Absolute pantheistic idealism was, in James's time, the dominant creed of spiritually minded philosophers. But James thought that the proofs they provide on its behalf were flawed (though personally I do not agree with him here so far as their essential core goes). It is simply a hypothesis with a certain amount going for it, but which has considerable difficulty in explaining those aspects of the world which are recalcitrant to spiritual values. Also it was even less able to cope satisfactorily with the problem of evil than orthodox theism, for the latter, in some versions at least, allows that God has produced a real other for himself in giving his creatures free will.

4. Finally, there was the hypothesis of a finite God. This offers an explanation of religious experience, of man's moral aspirations, and of psychical phenomena. It does not offer any final explanation of the being of things, for God is conceived, not as the ultimate explanation of everything, but just as one particularly important one among the various competing forces in the world. But then, according to James, none of the other theories offers any more successful answer to the question of why there is anything at all rather than nothing, so this is not really any special disadvantage on its part. It must be judged like the others as an account of how things are rather than as an ultimate explanation of why they are so.

As such, its great advantage over both orthodoxy and spiritual monism is that it avoids the problem of evil and does not threaten the sense we have of our own free agency. Moreover, it incorporates the scientific picture of things which led to naturalism, and yet interprets man's religious experience in the way it itself suggests. True, mystics usually interpret their experience monistically, but there is no real ground for thinking that the source of their saving experiences comes from a God either identical with, or the source of, the universe as a whole. They may just as well come from a reality which is the preeminent spiritual force within it.

James held that this conception of God and how man stands to Him was implicitly that of the ordinary religious believer who thinks that he meets God in prayer, that he can join with Him in fighting the good fight to improve the world (a fight in which setbacks, even if they can be overcome are not merely illusory for things do really happen which God would have preferred should not) and that it is by his own free will, (and not as an eternally fated part of God's plan) that he himself does his part well or ill.

Actually, James distinguishes two version of the hypothesis of the finite God. On the first we can feel confident that good is more than likely to triumph in the world, especially if sufficient good people lend God their wholehearted support, but cannot utterly rule out the possibility that things will end in total wreckage. On the second, God is bound to win in the end, but may do so more slowly and perhaps less fully, if finite moral agents choose wrongly. On this second view, God is like a great chess player who does not settle precisely how the game will develop, because that depends on his opponent as well as on himself, yet is sure to respond to whatever moves are made in a way which will lead to his ultimate victory. He thought his father had been more successful than most in synthesizing this view with a more monistic one. See the introduction to *The Literary Remains* of his father.[14] This introduction also contains William James's fullest discussion ever of the doctrine of the creation, though nothing very positive is said. Upon the whole his finite God does not seem to be postulated as a creator, and to that extent surely departs from anything which can be called ordinary theism. But James seems right that the conception is very much that of at least many ordinary Christian believers, so far as it is the conception of the state of things after creation and of their own present relation to God which is at issue.

(James has been accused of inconsistency in letting his conception of a finite God sometimes go with confidence in a final victory of good. But I do not see that this is so. James hesitates between the two views just described, but each is consistent in itself, and each leaves us humans with real free will and a real power to forward or retard a possible or certain triumph of good and to affect the form of that triumph if it comes.)

If, now, we use our will to believe to choose between these four views, or between the three which remain as serious options after the first has been discounted, James has no doubt that the preference should be given to the doctrine of the finite God. It is preferable to naturalism whose disheartening picture of the world is likely to sap our moral energies and depress our hopes of happiness for ourselves and

[14] Reprinted in James, 1982.

others. For not only does naturalism rule out any hope of an after life (though this is not of prime importance for James) but it suggests that humanity itself must come to an end with the destruction of the earth in a final cataclysm (James, 1937, pp. 103–7). It also suggests that free will is an illusion and that our higher aspirations are always liable to be thwarted by chemical changes in our body. And it makes quite illusory that sense of an encounter with what is most significant in the world which men feel in the highest experiences of which they know.

The doctrine of a finite God is also preferable emotionally and volitionally to spiritual monism.

The human value of spiritual monism or absolute idealism, both in its more academic and in its more popular forms, lies in the peace it can bring to our minds, when we are weary with our struggles, through its assurance that all is always eternally all right. Thereby it provides us with what James notoriously called a 'moral holiday'. But these, if too prolonged, risk damaging our moral sense and urge to put things right by encouraging us to think of the evils of the world as somehow essential elements in its total goodness as the life of absolute spirit.

Moreover, James thinks that most of us will feel that there is something rather effete in wanting a guarantee that things will go well. The sense that they will go well only if we and others like us lend a hand is far more to the satisfaction of most of us who want to play a genuine role in the world's affairs.

Thus the doctrine of the finite God has most of the advantages of the other two religious conceptions without their disadvantages. If it fails to give us an absolute guarantee that things will go well, that is something most of us do not want. And it is not as though it gives us none of the emotional support of absolute idealism, for it leaves us with the sense of something greater than ourselves on whose help we can depend if we seek it. It is thus a hopeful view of the universe, but one which avoids a specious apology for its imperfections. It is melioristic and progressive; it does not bid us blind ourselves to the badness of the world but it does give us hope that we can bring it steadily nearer to perfection. James's belief in the reality of time and Bradley's in its unreality are significant in this connection. For James the reality of time means that evil may one day be overcome and be as though it had never been, though its present reality condemns the optimism of the absolute idealist. For Bradley, in contrast, the universe consists ultimately of all times held together in the standing now of the absolute experience. On such a view evil is eternally there and cannot cease to be, and can only be redeemed, if at all, in the one eternal now.

So James recommends his conception of a finite God as on a par evidentially with the others and as preferable on passional grounds.

Though belief in it must rest partly on faith or the will to believe it is not evidentially inferior to the alternatives.

IV. Concluding Comparison

All in all, the differences between James's and Bradley's views on religion provide a paradigm example of James's contrast between the two types of supernaturalism. For Bradley, religion had no concern with the supernatural, understood as something abnormal sometimes intruding into daily life; its concern was, rather, with that absolute reality to which any phenomena whatever must belong. For James, in contrast, a religion not concerned with something that might thus intrude, was empty. True, they both thought religion more a matter of conduct than of belief, but for Bradley it was in the light of the nature which any reality must necessarily possess, while for James it was conduct in response to influences from a special realm.

Upon the whole my own sympathies are more with Bradley here. The idea that the great questions of how to live turn to any considerable extent on obscure historical facts or even on whether we will ourselves later live in a world beyond this one seems as wrong-headed as he thought it, and he is right, surely, that beauty, truth and goodness are as valuable in this world as they could be in any other.[15]

But I would also side with Bradley against more recent refined supernaturalists, if I can so call such thinkers of our own time as Don Cupitt, against a totally naturalistic picture of the world. I do not think there can be an adequate conception of human good if the world is conceived in purely naturalistic, at least if that means materialistic terms. But the alternative to materialism is not crass supernaturalism but a spiritual interpretation of all phenomena, both those which would commonly be called natural, and those which, if they occur, would commonly be called supernatural.

So perhaps we should distinguish three, rather than two, types of supernaturalism. Piecemeal or crass supernaturalism is a position which, though it covers many different outlooks, can stand as a general characterization just as James gives it, but refined supernaturalism divides into two types, which we might call scientific on the one hand, and mystical or metaphysical on the other. The first effectively supposes that the objective truth about the world is purely naturalistic in the sense that there is nothing in it or to it which cannot be characterized and explained in terms of the natural sciences. Its supernatural component only comes in as the apparent subject matter of a kind of

[15] Compare Don Cupitt, 1980, p. 41, also p.10.

non-scientific language game, that of religion, which fulfils a vital function for those parts of the natural world called human beings.[16] This language game, properly understood, is not a rival account of how things are to the scientific one, for it is not really an account of how things are or are to be explained at all. It is rather a certain kind of performance, that of telling morally invigorating fables and engaging in ceremonies of a psychologically inspiring kind.[17] Matthew Arnold was the nineteenth century figure who most nearly approximated to this type of refined supernaturalism, and much was he mocked by Bradley for doing so (Bradley, 1927, pp. 317–18).

In contrast, the second type of refined supernaturalism, the mystical or metaphysical type, is a rival to the scientific account of the world, not because it refers to particular occurrences and phenomena which somehow escape scientific treatment, but because it thinks that the world has a pervasively spiritual character whose presence has nothing to do with the amenability of any of its details to scientific treatment, and that this pervasive character of the world calls for and makes possible moral and emotional attitudes which cannot be grounded in anything recognized on a purely scientific account.

Such a pervasive character of the world may reveal itself especially to certain states of mind, whether mystical, ethical, aesthetic or otherwise, but what is revealed to these is essentially an aspect of everything and discoverable in principle everywhere and everywhen. Empiricists and pragmatists of a certain type are likely to identify the two types of refined supernaturalism, because the spiritual dimension posited in the second type of refined supernaturalism seems to them vacuous unless interpreted as non-cognitive performances. But in this I believe such pragmatism and empiricism to be mistaken.

[16] This may not be how Cupitt would characterize his own position but it is what it comes to. For while he sometimes talks as though there were no objective truth at all, what he seems to mean is that there *is* just what is properly called the natural world, though humans can only get to it through culturally relative symbolic systems. For his commitment to what is really the naturalistic conception of the world peculiar to his own culture (not for that reason, in my view, necessarily failing to be objectively true) masquerading as the refusal to 'privilege' one culture over another see Cupitt, 1989, at p. 68 and passim. At times in his discussion of truth and in his insistence that there is nothing but the human world of signs (see p. 142), Cupitt indeed sounds quite like James but James had a genuinely open mind about how things may really be quite lacking in Cupitt for whom the sole reality of the world acknowledged by the common or garden twentieth century European intellectual is a pervasive dogma.

[17] Santayana was a rather different sort of refined supernaturalist of the scientific sort. See especially Santayana, 1905, 1940 and 1946.

References

Bjork, Daniel W. 1983. *The Compromised Scientist, William James in the Development of American Psychology* (New York: Columbia University Press).

Bosanquet, Bernard. 1889. *Essays and Addresses*. London.

Bradley, F. H. 1969. *Collected Essays* (Oxford: Clarendon Press). First publ. 1935.

Bradley, F. H. 1927. *Ethical Studies*. Oxford: Clarendon Press. First publ. 1876.

Bradley, F. H. 1930. *Appearance and Reality. A Metaphysical Essay* (Oxford: Clarendon Press). First publ. 1893 and 1897.

Bradley, F. H. 1968. *Essays in Truth and Reality* (Oxford: Clarendon Press). First publ. 1914.

Cupitt, Don. 1989. *Radicals and the Future of the Church*. London: S.C.M. Press.

Cupitt, Don. 1980. *Taking Leave of God*. London.

James, William 'The Will to Believe', in James, 1923.

James, William. 1909. *A Pluralistic Universe* (New York: Longmans, Green and Co).

James, William. 1982. *Essays in Religion and Morality* (Cambridge, Mass.: Harvard University Press).

James, William. 1924. *Memories and Studies* (New York: Longmans, Green and Co.) First publ. 1911.

James, William. 1937. *Pragmatism; A New Name for Some Old Ways of Thinking* (New York: Longmans, Green and Co). First publ. 1907.

James, William. 1928. *Some Problems of Philosophy. A Beginning of an Introduction to Philosophy* (New York: Longmans, Green and Co). First publ. 1911.

James, William. 1975. *The Meaning of Truth* (Cambridge, Mass.: Harvard University Press). First publ. 1909.

James, William. 1923. *The Will to Believe and Other Essays in Popular Philosophy* (New York: Longmans, Green and Co). First publ. 1897.

James, William. 1985. *The Varieties of Religious Experience. A Study in Human Nature*, ed. John J. McDermott. (Harmondsworth: Penguin Books Ltd). First publ. 1902.

Jones, Kelvin I. 1989. *Conan Doyle and the Spirits*. The Aquarian Press.

Perry, R. B. 1935. *The Thought and Character of William James. As revealed in unpublished correspondence and notes, together with his published writings* (Boston: Little Brown and Company). Two vols.

Santayana, G. 1946. *The Idea of Christ in the Gospels* (New York: Charles Scribner's Sons).

Santayana G. 1905. *Reason in Religion* (New York: Charles Scribner's Sons).

Santayana G. 1940. *The Realm of Spirit* (New York: Charles Scribner's Sons).

Religious Imagination

RONALD W. HEPBURN

I

In some recent theological writing, imagination is presented as a power of the mind with crucial importance for religion, but one whose role has often suffered neglect. Its fuller acknowledgment has become a live issue today. 'Theologians', wrote Professor J. P. Mackey, 'have recently taken to symbol and metaphor, poetry and story, with an enthusiasm which contrasts very strikingly with their all-but-recent avoidance of such matters' (1986, p. 1). As well as relevant writings by Eliade and Ricoeur, there have been treatments of religious imagination by Professor John McIntyre in his *Faith, Theology and Imagination* (1987) and in J. P. Mackey's composite volume entitled *Religious Imagination*.

What is beyond all question is that in the field of religion, imagination must be accorded an enormous role, seen as an indispensable agency without which the claims and teachings of religion could never be communicated at all—far less arrestingly or memorably expressed. It is imagination that carries the worshipper from the crucifix held in his hand, from the icon on the wall, from the fragment of bread and the sip of wine—to the thought of salvific events and transactions, on a cosmic scale and with a cosmos-transcending being. Imagination can, nevertheless, also be an all too willing religious worker, too ready to leap abysses in understanding and argumentation over contestable religious concepts and claims. It can be too ready to embrace lovingly a *prima facie* contradiction, and to give assurance that attributes of deity which seem quite incompatible are 'ultimately' compatible—perhaps becoming so at an infinite (but not thereby prohibitive!) distance.

I wish to attempt three things in this essay: first to offer some personal comments on the work of religious imagination; second to attempt some critical appraisal of a small sample of those recent studies, and lastly to raise certain fundamental questions about the validation of religious images and symbols.

II

Any account of religious imagination must start by acknowledging the role imagination as such plays in the very construction of our perceived

Ronald W. Hepburn

world, the *Lebenswelt*. Imagination strives to impart structure, to synthesize, restlessly to go beyond, to transcend its own syntheses at every achieved level. Intensifications of these operations quickly carry us towards its distinctively religious activity. The absolutely basic—and I shall adopt a roughly Kantian-Strawsonian account—is imagination's epistemological role in converting undifferentiated sensation into a world over-against myself as a subject, and my fashioning a conception of subjecthood through my interaction with what I come to understand as an other. It is not in terms of any new ingredient in experience that those basic 'conversions' are constantly achieved, but by acts of imaginative grasp and synthesis. If, for instance, I am to distinguish, out of the manifold of my sensations, a world of objects external to myself, not merely elements in a phantasmagoria, I must imagine at any time what is necessarily not given at that time—hidden sides, alternative perspectives, continuity of the world and its objects beyond the immediately sensed; I have to imagine pasts of objects and futures for them which are not just the past and future of my sensations. The hidden may come to perceptual light, alternative perspectives be sampled, the enduring of objects lived through. But, the vital first contribution of imagination is to insist: 'this mass of sensation is not all—is not exhaustive'. There is an independent reality, an 'over-against'. So a 'space' is interposed by imagination between my global sense-awareness and a world I can now speak of 'perceiving'. With our knowledge of other persons: though we are confronted with a continuity of perceptible change, movement of bodies, limbs and faces, imagination will not take that as exhaustive of this personal reality, but goes beyond what already might seem a *plenum*, to posit centres of consciousness, will, intention, reason.

In these roles imagination does not anticipate experience in the sense of suggesting, proposing what in more favoured, privileged conditions could be subsequently confirmed or rebutted by non-imaginative means. We cannot coherently conceive of an improved mode of perception of external objects or a more intimate access to other minds in which imagination's part is superseded and done away with. It must remain an essential element in cognition: and that 'must' can be called 'transcendental' in a strictly and strongly Kantian sense. I perceive a world *now* and with *this* sense-manifold, only if imagination posits a multitude of actualities, pasts and futures which are *not* part of my momentary experience. As Strawson put it, such non-actual perceptions are '*alive* in the present perception' and that present perception (to *be* perception of objects at all) is 'soaked with', 'animated by', 'infused with . . . the thought of other past or possible perceptions' (1974, pp. 53, 56f).

Central and obvious is imagination's power to 'envisage a state of affairs different from that of which we are aware' (Sparshott, 1990, pp. 4f), to envisage 'alternative orders', alternative worlds; and such extensions readily take us into imagination's inventive, creative roles and to the more daring instances of its essential work of 'going beyond'. Without it we should be able neither to form concepts nor to apply them in particular cases. But it 'goes beyond' that role also. In contexts both of art and religion it seeks to go beyond concepts so as to mediate experience that defies conceptual analysis. As Mircea Eliade wrote in his *Images and Symbols*, 'the power and the mission of the Images is to *show* all that remains refractory in the concept' (1952, 1961, p. 20).

Finally, in its insatiable *nisus* for going beyond—'transcending'— imagination does not draw back from seeking to transcend the entire phenomenal world, the world of lived experience: at the very least to animate and keep alive the thought that, although such transcendence is literally and necessarily inconceivable, it is nevertheless an insuppressible extension of imagination's concept-transcendence and its other even more fundamental and familiar transcendings of level, from sensation to world of experience, from behaviour to others' minds. Wonderment at the sheer existence of a world can tip over into a kind of incredulity that it exists 'on its own'. The experience I am thinking of is the work of imagination urging its way 'beyond'. Equally insuppressible, to the religious mind, is its synthesizing, unifying drive, whereby it seeks to interpret sporadic 'visionary moments', 'spots of time', fugitive nature-mystical experiences, as all brief liftings of the veil upon a single divine transcendence, part-disclosures of a single mystery.

Could it not be argued, however, that this account so far has been excessively optimistic about the religious relevance and constructive efficacy of imagination? Should not some critical queries and caveats be interpolated before we go farther? Over-optimistic, yes, in certain obvious respects. Transcendental arguments can secure and validate imagination's activity (its going beyond) in constructing the world of things and persons, but I see no parallel transcendental argument necessitating or justifying the leap from world to a divine, transcendent ground. Imagination prompts, lures, urges; but can do no more . . .

Suppose we point to the 'leap' we make, from the movements, the behaviour of others, to the minds of others. The cases, again, are very far from parallel; since in order to reach the level of self-awareness, to have the concepts and language necessary for raising questions about imagination and its possible leaps, we must already have acknowledged the reality of other minds. We can, in thought, refuse to transcend world, to God, without comparable incoherence. Yet again, when our symbols, however potent and evocative they are, *conflict* with one

Ronald W. Hepburn

another, imagination alone cannot resolve the conflict. We can easily come to expect too much of it. The Romantics certainly did so, and a few of the recent writers on religious imagination come near to doing so again. It is all the more important to map the scope, and to raise questions about the limits, of imagination, of symbolism and myth and their varying relation to truth. It is not on account of blunders, over-sights and muddles that theories of imagination have, over the centuries, oscillated between the denigratory and the deifying, have cast imagination now as hero, now as epistemological villain, and even within a single theory have shown deep ambivalence.

Pascal, in the *Pensées*, wrote memorably of the *grandeur et misère*, *grandeur et faiblesse* of mankind: the co-presence in man of paradoxical qualities (1670. Article VI *passim* and particularly para. 416). The phrase keeps suggesting itself to me as applying very powerfully, within human nature, to imagination itself. As well as being vitally active in giving structure to the real world, imagination can obviously draw us away from it. In philosophy, for instance, it can fall under justified suspicion of doing that. At the end of his well-known paper, 'Imagination and the Self', Bernard Williams writes: 'At least with regard to the self, the imagination is too tricky a thing to provide a reliable road to the comprehension of what is logically possible' (1976, p. 45). And we can turn to Kant for some particularly apt characterizations of imagination's greatness and littleness. Imagination is a 'blind but indispensable function of the soul' (1781, A78/B103). 'In lawless freedom imagination, with all its wealth, produces nothing but nonsense; the power of judgment, on the other hand, is the faculty that makes it consonant with understanding.' Kant can speak of 'the might of imagination'; but he can also offer a theory of the sublime in which it is crucially imagination that is—not triumphant but—humiliated, overwhelmed by great magnitudes or energies of nature: and it is through the contrast between awareness of imagination overwhelmed and awareness of reason and the free moral self, which are by no means overwhelmed, that the oscillatory, dual experience of fearful exhilaration (sublimity) is generated and sustained (1790, paras. 50 & 26; paras. 23ff).

For a more basic instance of the present theme, we can go back to the role of imagination in the first construction of the phenomenal world. That interposing, by imagination, of what I metaphorically termed a 'space' between an immediate sense-manifold and what thereby becomes for me a world, that interposition has at once its immense *grandeur* and its nullity, the quality of a *néant*, dimensionless and contentless. It brings the bare thought: 'this sensation-mass is not all, does not exhaust what is'. In this instance, we can say that the greatness lies precisely in the littleness—*multum in parvo*.

The paradoxes and oxymorons certainly persist within the specifically religious employment of imagination. Imagination is only too willing to offer images of deity, but each has to be negated as it appears—as anthropomorphic and so demeaning to the divine. If infinite and transcendent, then not imaginable: whatever we *are* able to imagine, and so grasp, cannot be the real object of our quest. There may well be a high religious importance in the very striving-and-negating; but illusion may be no less at work, and we shall not determine *which* by still further appeals to religious imagination. That is to say: if imagination claims success here (imaging the divine), it betrays its failure; if it strenuously endeavours, then confesses its failure, it *may* be nearer to success, though it cannot itself tell us if this is really so. We cannot infer from confessed *un*-imaginability to the actuality of an imagination-transcending being. Nor can imagination alone tell us how close or remote are its conceptions, in relation to the Reality which it takes as its intentional object.

What option we take at this point of our exploration will inevitably depend on our prior religious commitment, if any; on our individual openness to religious experience, and on the interplay between that experience and our metaphysical reflection, our sense of reality at any time. What is so elusive, we may say, so questionable in its status, what cannot be 'cashed', made determinate, imaginable, is better abandoned to naturalistic-reductive explanation in psychological and sociological terms.

As another alternative—religious–agnostic, we could call it—we may opt to live with the ambiguities, and to allow religious imagination (perhaps encourage it) to extrapolate endlessly, to go on making its transcending gesture, although unable, necessarily unable, ever to translate that into any single, determinate, spelled-out 'message' or dogma, which we would not immediately know to be deceptive.

Many writers in the religious traditions would say that the ambiguities come precisely from neglecting the role of *revelation* in any knowledge of God we may have. The images—the system of images—that should most concern us are *revealed* images, without which we cannot speculate or extrapolate our own way to the transcendent. *With* them, and only with them, we are furnished with an indispensable, indirect but wholly adequate, knowledge of God, and the way of salvation. Because the images, symbols, myths are God's only way of addressing us, we are released from the impossible task of tailoring them to the infinite and eliminating the anthropomorphic.

Attractive, and of course theologically prestigious though this position must be, it is by no means free of metaphysical and epistemological problems. Though the content of an alleged revelation may be carried in parable, myth and image, the claim that what is so carried is indeed God's self-revelation to man can be authenticated only if we have a convincing account of God as its revealer, an account distinct from the set of images we are to contemplate as revealed. We need a coherent concept, of the *transaction*, God to man, which constitutes the imparting of the revelation and can alone allow us meaningfully to take the images in this way. In epistemological terms, we need the means by which to make intelligible the referring of an image-system, to a divine source.

(I acknowledge that this thought-model, though indispensable, is still over-simple. In section IV I argue that imagination may well be involved also in trying to articulate that latter task—of referring to God: though not in its image-contemplating mode.)

Rather than develop these arguments about revelation here, I want to return to the reflections that preceded them. Suppose there is no vindicating of imagination's visionary glimpses, its stammering after transcendence, and yet we are deeply unwilling to relinquish them to reductive naturalistic explanation. What then?

From certain points of view that situation can present itself not as a spiritual defeat but as a moment in a still spiritual journey. It is not just that we may find ourselves to lack intellectual resources to determine whether there exists a God answering to our clear concept of such a being, or a transcendent realm answering to our clear concept of that, and so on. Our problem (though it is not problem only) is deeper. We have (as I said) to negate every image of deity that imagination proffers to us; and the same will be true also of every image of a hereafter in the presence of God. Imagination tries to bring into sharp focus, to make determinate, its visionary extrapolations; but so far as it supposes itself to have done so, it at once knows it has *falsified* its original vision. Not only falsified but trivialized its vision: for it belongs to the logic of perfection that to falsify must be to trivialize. The experiencer can be tempted inwardly to substitute the trivial and falsifying 'story' or dogmatic 'message' for his scarcely recollectable surmise or mystical experience or vision. He may accept that invitation, but be troubled, left with a sense of being in *mauvaise foi*. It is as if we had been deeply moved by a melody, and were later to be perplexed and disappointed when we heard it again or played it at the piano. In the end, we realize that what had made it banal and trivial was the substitution of a trite harmonization for the authentic and 'magical' version.

To hold to, not to betray, the unconceptualizable, unimageable transfigurations of experience, neither forcing them into alien moulds nor ruthlessly rejecting them: this can be seen as faithfulness to an inner religious logic, not an expression of scepticism. It is the logic that negates all substantializing and localizing of transcendence, all repetition in the transcendent of the concepts and categories of the life-world. Perhaps it urges us instead to hold to, to stay with, the strange vitality of the symbol or (as Karl Jaspers (1932) would have put it) the cipher. It should be added that, minimal though its metaphysical 'exposure' may be, if such an undogmatic, religious–agnostic faith is to witness to anything at all—if it is to have any content—it can have no claim to exemption from philosophical *criticism*, any more than has the revelation-centred theology of traditional Christian belief.

III

Some writers who seek to support a specifically Christian view invoke imagination as a main cognitive instrument. Others altogether renounce theistic doctrine concerning transcendence and give over the field of religion entirely to imagination, to poetry and myth. I should like now briefly to sample both of these and to identify some critical challenges to which they may be open. I shall start with some topics from Professor John McIntyre's *Faith, Theology and Imagination* (1987), and his chapter 'New Help from Kant', in *Religious Imagination* (1986).

In the latter, McIntyre makes use of Kant's conception of 'schema', applying this to religious, parabolic thinking. In the teaching of Jesus, parable is a 'procedure . . . for the imaginative proliferation of images which represent a certain concept', most often that of the Kingdom of God. Again, as in Kant, 'imagination synthesizes the manifold of the intuitions', so faith may be in part understood 'as religious unity of apperception'. Faith conditions 'a unity in the consciousness of the believer', *a priori*—a 'transcendental' role. Faith, so construed, in turn provides the 'unifying condition' for an 'overview' of our empirical existence. As Kant inquired into the conditions of the possibility of experience (of an external world, of moral obligation), so we may seek the

> conditions of the possibility of certain people speaking of a God who for them has an independent and continuous existence, speaking to such a God in prayer and worship, and generally living their lives in the context of the reality of that God's being. (1986, pp. 106, 114, 115f, 112)

Much of the account John McIntyre offers of imagination's role in religious thought and discourse is illuminating and incontestable. I shall comment, however, on his just-mentioned extension of Kantian

Ronald W. Hepburn

transcendental argumentation. When McIntyre refers to Kant's question about the conditions for explaining and sustaining the moral imperative, he allows that a-moral people may 'disclaim awareness' of it; and accepts that non-religious people may similarly not acknowledge any religious experiences whose conditions-of-possibility require to be investigated in their case. He admits that religious experience is 'open to other interpretations' and has no immunity from critical attack (1986, pp. 112f). This is admirable caution, but the outcome is an account without any of the rigour promised by the phrases, 'transcendental argument' and 'transcendental method'. Returning to the quoted sentence in the previous paragraph: should we not say that 'the condition of the possibility' of people speaking of and to a God 'who for them has an independent and continuous existence' is simply the belief that such a God exists and has the qualities that evoke this worship—a belief that may be true or may be false. I could not in similar style affirm that I may have no freedom or spontaneity of mind (conditions of rational deliberation and affirmation), without thereby undermining my entitlement to make that very claim—as a rationally argued conclusion. It would be wiser to reject Kant's own appeal to transcendental argumentation in the distinctively moral sphere than to blur the boundary between cases to which such argument applies in its full strength and those to which it does not.

In *Faith*, *Theology and Imagination*, John McIntyre argues that there are numerous central themes in Christian thought where the work of imagination is vital:—in God's own daringly imaginative act of sending his Son; in Jesus' teaching; in the images whereby alone we can interpret the Atonement (1987, pp. 54–61). Let us pause there. I think that the account McIntyre offers of images in relation to Atonement is both arresting and problematic. Among these images are 'ransom', 'reconciliation', 'sacrifice', 'propitiation'. Despite theoretical conceptualization, this 'vast range of images' has 'refused stubbornly to go away'. Other images play a 'relating' role: 'substitute', 'representative' and 'on our behalf'. It becomes clear that to conceptualize here is also to show the 'incompleteness' of these images or symbols. To whom, e.g., is the ransom to be paid? The model of 'reconciliation' works in reverse of expectation and logic, for God reconciles *us* to himself. 'Sacrifice', but not to a God in need of appeasement (and similarly with the image of atonement itself): 'propitiation' but not to an angry potentate . . . All such are 'incomplete symbols'; 'they do not follow their own structures to their rigidly logical conclusions'. But it is only through such images that 'we come to forgiveness': they are 'the paths to salvation'. The doctrine of the atonement lay 'too close to the heart of Christianity to be conceptualized into ineffectuality' (1987, pp. 59, 56f, 60f).

Perhaps John McIntyre is right. Yet, from outside, his methodology has disquieting features. It is admitted that, powerful though they are, the images are fragmentary, one might say, 'broken'. Their brokenness is both to be acknowledged and accepted. It is made less questionable by the technical description, 'incomplete symbol'. What is disquieting is that such a move would seem to give a licence to any theologian or ideologue of any belief-system to disregard logical incoherences, so long as some image or symbol carried, as *one* of its aspects, an implication for appropriately regulating our attitudes, evaluations and actions. Allow that, however,—generalize it—and all discriminatory power is lost. Appeal to 'imagination' could (to put it crudely) get a theologian off all hooks!

McIntyre does not generalize. He sees himself as dealing with images which are an important part of distinctively Christian truth. His fundamental commitment to a Christian perspective lets him see the structure of images or 'models' as a mode of revelation. In his words, 'they represent the processes initiated by God to give effect to forgiveness' (1987, p. 60). *Without* that commitment, one might equally well see the 'incompleteness'—the brokenness as I am calling it—as ground for *rejecting* these images: they do not really fit, when you think them through. We might then go from the illogic of the images to the incoherence or at least unsupportedness of the doctrine, to the extent that the doctrine depends upon them. Or again: less radically, one can argue that, in view of the incompleteness of many of the images of Atonement, the main weight in a philosophical theology cannot fall on these, but rather on whatever it is that prompts us to accept the images, cling to them, in spite of their incompleteness or brokenness. And that, presumably, whether experiential or rational, will not be simply a matter of entertaining further images or symbols—least of all, symbols that are themselves incomplete.

Clearly, for most Christian theologians, the ultimate account will be in terms of revelation: the images proceed from God, and have their authority from God. That is why their incompleteness is seen as no ground for rejecting them. Yet, although McIntyre and others have called that overarching account itself a 'model'—the model of revelation—to include it in this way among the images, symbols and models will not provide that ground for retaining incomplete symbols which we need.

However revelation is imparted, the response may come, we are not entitled to expect a perspicuous, non-mysterious account in which both ends of the relationship, divine and human, are simultaneously in clear focus, together with the line of communication between them. That schematic thought cannot fail to imply that God is an empirically knowable being among beings; and theologians will remind us that since he is a 'hidden God', to expect a direct account is to ask the impossible.

That would be an important and understandable move. The theme of a hidden God itself suggests great scope for religious imagination, filling out our surmisings, 'glimpses' of the divine. Indeed, the more hidden, the more work for imagination to do. Nevertheless, the epistemological danger is that beyond a certain degree of hiddenness, God must begin to lose any features which alone can entitle a theist to speak of him as actual, rather than imaginary in an unacceptable, fictive sense. The belief that God is both actual and hidden in fact keeps its meaningfulness only so long as we can conceive, or imagine, God as *less* hidden, and extrapolate some way, even if 'darkly', towards 'God hidden no longer': only if we can envisage an epistemological improvement. Roughly speaking: the concept of a hidden God is dependent, or parasitic, upon the concept of a non-hidden God. We cannot logically be thoroughly agnostic about the second, while being (also thoroughly) wedded to the first.

The problem of controlling, referring, connecting the set of images or symbols, at one end to God and at the other to man, naturally ceases to exist as a problem, if we simply give up the attempt to 'attach' the images to a self-revealing transcendent deity. And this is by now a familiar enough theological 'revision'. By means of it even more work is secured for imagination, as talk about deity, transcendence, revelation themselves are (wittingly) internalized within the structure of image and myth.

Two very different roles, or epistemological contexts, are thus identified for religious imagination. In the one (traditional Christian), imagination provides a partial, but reliable, access to the ultimate truth—now through a glass, but with the expectation, eschatologically, of *emerging* to a more direct vision: no longer by faith but by sight. The images are to be superseded, but not discredited. In the other, if it is consistently thought-through, we accept that there will be no emerging from the images: we live with them and by them, and will not experience, eschatologically, their falsification . . . nor their verification either. Add to these a third option: that we lack the means to decide between the first two: a third option with considerable attractiveness. To get oneself into the position of discriminating most reasonably between the alternatives, one must already have found at least tentative answers to fundamental problems in philosophy of religion: problems in which imagination is again doubtless involved!

If an actual divine origin for the images is, in fact, deleted, the images become a 'background', prompting, regulating and meditatively sustaining the attitudes and evaluations that now constitute the religious orientation. They can perform these tasks even without being fully analysable into a coherent, logically compossible set, so removing another of our anxieties.

But a new decision lies ahead. Within such an option, ought we to press on towards demythologizing the images and symbols; or should we heed those who urge us to hold to the images, to stay with the language and modes of thought proper to myth? The demythologizer believes he has reached bed-rock only in a set of concepts of a broadly existentialist kind—authenticity, self-scrutiny, integrity, *agape*, the avoidance of self-deception. To his critic, however, this is a misguided attempt at reduction: misguided because myths are irreducible, their truth 'cannot be decoded'. I am quoting from one such critic, Colin Falck, in his *Myth, Truth and Literature* (1989, p. 130). The meaning of myths is a part of our vision of the world: we have a lasting need for myth, and that need must manifest itself today as the task of *re*-mythologizing rather than of deleting imagination's contribution so as to leave us with only a 'stripped and bare landscape'. (Ibid., p. 127: Colin Falck is alluding to Don Cupitt.) Symbols, images, myths, then, are neither to be given a dogmatic privilege, as by the institutionalized world-religions, nor seen as merely transitional, awaiting adequate paraphrase. Paraphrase is heresy in poetry, and religion (dogma apart) *is* poetry. 'Poetry replaces religion.' Not that this is seen as an abandonment of concern for *truth*: 'literature . . . gives us our purest and most essential way of grasping reality or truth'. 'Literature remains the most reliable access to reality that we can have.' The poet 'is a sage', Falck claims, quoting Keats, 'A humanist, physician to all men'. Like many other writers on religious imagination, some less radically revisionary than himself (e.g., Eliade, Dillistone, Jung, Berdyaev, Durand), he sees a need for the redeeming of the 'mechanised and de-sacralised world of practical life', one compatible with a claim like William Blake's that 'all deities reside in the human breast' (Falck, 1989, pp. 127, xii, 3, 169: Blake, 'The Marriage of Heaven and Hell', Plate 11).

With this last aspiration I have every sympathy. At other points, though, I am uneasy over Falck's programme of a thoroughgoing reinterpretation of religion—certainly where it includes Christian religion—in imaginative–poetic terms.

First, he is hostile to the giving of privileged authority to 'dogma', while claiming that myth and literature (which can and should assimilate religion) are themselves concerned with truth. But surely, except where it is arbitrary and unsupported, the dogmatic element in Christian thought sees itself as no less truth-concerned. It is misleading to say that minus the dogma poetry and religion are one and the same. The 'literary criticism' which is to engulf theology, and the art which is to be 'a counter to dogmatic belief' (1989, p. 169) can hardly be competent to appraise the complex cosmological–metaphysical and the highly specialist historical components in the Christian truth-claims. A decision has already been implicitly taken that these claims are *false* or at best poorly grounded: and the extent to which a Christian orientation must be *response* to belief in their truth is insufficiently understood or acknowledged—as it is also by the demythologizing revisionary theology.

If myths cannot in their nature be 'decoded', we may see ourselves as, strictly, prevented from even raising the question whether components in the myth do or do not refer truthfully to the world: in that way, questions of truth could become confined (surely unwarrantably) to questions of faithful expressiveness of human *feeling* about the world, truthfully or falsely understood.

If we understand religion as essentially a structure of myth, and categorize it as such along with poetry, literature, the arts in general, we thereby give the poet, and his interpreter and appraiser the literary critic, the status of sage. Falck, we saw, accepts this, and invokes Keats to support him. Certainly, the insights of poets should be fully acknowledged and reckoned with, but their authority can also be easily exaggerated; there is still serious point in Plato's ambivalence about the poets. Philosophy, as well as poetry and criticism, must continue to have a voice in questions both of fact and the interpretation of fact; and values need to be argued over as well as expressed and celebrated. Metaphysical and logical appraisal cannot be relegated to the margins.

Falck himself does not escape the need for these: he adopts, as we noted, Blake's thoroughly metaphysical view that 'all deities reside in the human breast'. That claim belongs to a 'metaphysical supplement' quite as much as does the traditionalist's story about revealed images proceeding from an actual deity. Certainly, the claim is made, by Blake, within a work of literature; but neither Blake's nor, clearly, Falck's understanding of the claim can be such as will confine it to its literary context: metaphysical it unquestionably is. Nor again can it be left simply as self-evident, or unsupported, undefended against its critics.

A rejoinder might be made on the following lines. The question about a religious image or myth is not whether it has metaphysical support or grounding, but whether it has a *hold on the imagination*. That is to say, we do not need to leave the sphere of the imagination and its workings. Consider Wittgenstein's well-known example of a powerful religious image—the Last Judgement. If this image has a hold on my imagination, I shall orientate my reflection and my practice, my self-monitoring, by reference to it. I rather think, however, that if I were to see it as no more than that—an image with a hold on my imagination, whose role it is to regulate my conduct and accountability—it would come to *lose* its imaginative hold. I can readily sympathize with Wittgenstein's sense of incongruity between belief in the Last Judgement and factual or scientific beliefs about future events. It is perfectly possible for me, nevertheless, to believe in a Last Judgement as something I shall experience (indeed, be unable to escape), even though I am altogether agnostic about its connection with the series of everyday events in time. No doubt it will be radically discontinuous with these—and maybe there will be no trumpet: yet I may be sure that there will be *more than images*. So the image, the picture, is not all: neither is the mere empirical fact that during some stretch of time the picture has a hold on my imagination. If there cannot coherently be more to the matter than 'picture', then again imagination shows itself to be too deceivable, too 'tricky', to be an adequate guide in this area.

IV

A constant feature of this study has been the tension between appeals to religious imagination as having a uniquely revealing role, on the one hand, and (on the other hand) critical arguments to the effect that a *further* appeal is indispensable—to a continuing metaphysical or logical, sometimes moral, *critique* of the images, and a rationally sustainable account of the connection between images, myths and God.

Given the deceptiveness of imagination, the incompatibilities among its products, the need for some 'meta'-account of its very claims to *be* the vehicle of revelation, one can hardly be judged misguided or perverse if one seeks ways of 'grounding' or of critically questioning claims about its authoritativeness. Surely a duality of a (schematically) Popperian type is indispensable: imagination speculates, with freedom and passion, but is necessarily checked and controlled by critical reason. Such a dual account can, admittedly, be damagingly sharp. J. P. Mackey is persuasive when he deplores the domineering, tyrannizing potential of reason—if given, in the end, a place of total ascendancy; and he contrasts that with the gentler, freer power of imagination to 'lure' and to 'haunt' (1986, pp. 16f).

Mackey's own way forward takes him to a concept of 'vision', which combines imaginative inspiring and sustaining, with reason's work of analysis and clarification, the figuring out of implications. Mackey concludes that the 'source-experience' of Christian faith is 'essentially imaginative': the question of its claims to truth involve, therefore, and crucially, how 'truth-claims of imagination' are to be understood and appraised.

Here Mackey appeals to the idea of 'truth-claims that can only be settled by action' and in terms of vitalizing transformations of human nature: notably through the eliciting of the 'most enhancing' passion, love. Jesus, he says, 'was not concerned with the existence of God', but with introducing 'the reign of God'. Imagination has the power to 'forge . . . unity of vision, passion and pattern of attitude and action'; and it is to all these that we should look in considering claims to truth (*ibid.* pp. 19–23).

I should like to offer two critical comments on this pragmatic account. First, the idea of 'vision' as ultimate here—holistic, unitary, is an attractive one; but surely there can be no guarantee that the rational component will confine itself to the working out of implications, analysing, clarifying in ways that support, reinforce, but never disturb or show up flaws in, the visionary component. If rational enquiry is going on—in any form—it must be capable of disturbing, disrupting, subverting, as well as confirming, the imagistic elements, through a critique of their implied affirmations and the presuppositions of these. We may naturally want to keep the visionary package wrapped and unopened, but reason, *qua* reason, can never promise to let us do so.

Secondly, the basic difficulty with any pragmatic approach is this: a vision may be well-unified, imaginatively and morally inspiring and supportive, and yet the whole system of ideas may be erected upon false presuppositions. These presuppositions may not be the believer's direct or passionate concern, but they are nevertheless necessary to the underpinning of the valued practical efficacy of the vision itself. (The vision may be of God's *reign*, but God's *existence* is presupposed in the symbolism of his reign.)

To claim that the view of the world that releases and harnesses human energies, vitality, spiritual potential—and to the highest degree—may nevertheless be false, must be unwelcome and hard to bear. But I can see no way of denying it with confidence, unless I accept the very bold presupposition that we know the universe to be such as dependably to conserve value, to foster and to maximize its realization. But that presupposition can hardly be affirmed without any grounding: it is surely the desired conclusion of religious enquiry rather than its starting-point! Nor, obviously, can it be seen as itself pragmatically justified. That would only push back a further stage the final validating of any and all pragmatic appeals.

To put the matter sharply and simply: we can be cheered and inspired by a belief that is false, and the truth can on occasion be dispiriting and life-impeding rather than life-enhancing. The pursuit of *truth* and the pursuit and enhancement of *life* are both fundamental value-pursuits, but the pursuit of irreducibly different values, and we cannot count on the first invariably to further the second.

It might be supposed that by this point I had convinced at least myself that religious imagination, though given the greatest scope, stands in need of a metaphysical regulator or supplement not itself imagistic, or mythological or symbolic; and that this schematically marks the boundaries or limits of imagination in this domain. I have certainly relied throughout on the possibility of making two distinctions. The first is the distinction between certain of the products of religious imagination such as images and myths, and the means by which they can be affirmed and shown as *referring* to, or revelatory of, a reality beyond themselves and indeed transcending the life-world. The latter referential meta-account, as I say, cannot be a further entertaining or contemplating of images and the like. The second distinction is between the language of imagination and the language of rational appraisal and criticism.

Now, it is not hard to see how these distinctions may be challenged. First because it is hard to find specimens of actual metaphysical writing that are not themselves 'imaginative'; *impossible* to find any, if (very reasonably) we include live *metaphor* as among the products of imagination. One has only to recall, for instance, the web of metaphor connecting spatio-temporal particulars to Plato's Forms, and bold metaphorical extensions of everyday (or specialist) concepts by many other metaphysicians so as to cover the cosmos as a whole—an organism, a machine, a work of art, the product of a *will* that is nobody's will . . . Nor do contemporary philosophers rely any less on metaphor. Does it make sense, therefore, to speak of referring the religious symbols and images to a divine 'source'—*non*-imaginatively? Likewise, the second distinction is challenged, since philosophical criticism tends no less to be couched in metaphorical-imaginative discourse. So: how do I respond?

If we were already theists, it would be easy to admit that of course the language that refers to, links our images to, deity, will itself be strained, will need uncommon resources, to do its work. We shall have to reach for further 'incomplete symbols' in the task of relating images to God. Certainly: if we already have reason to say that we are dealing with, talking about, an actual deity . . . But if we are not in that position, not already convinced of the truth of the theism which we are enquiring into, we cannot appeal to the authority of revealed imagery in order to hold at bay our puzzlement at, or tentatively sceptical rejection of, the images *as* incomplete or broken or paradoxical. In their meta-account, some theologians do think of their key concept of revelation as itself a 'model', a further metaphor. Are we not, then, rendered unable to effect any rational and critical appraisal of religious images and metaphors by the fact that no wholly non-metaphorical idiom seems available, or perhaps possible? and that therefore there can be no 'stepping outside' the metaphors and models to a point of vantage from which they can be rationally appraised?

The problem is real, but not insuperable. For a start, we do have to grant that if all live metaphorical discourse involves imagination, there can indeed be no move to an imagination-free zone from which to appraise the products of imagination. None the less, I think we can make various distinctions *within* thought and discourse which do involve imagination, distinctions between different sorts of work, different sorts of operations we are seeking to carry out in different contexts.

Clearly, not all imaginative activity—even when concerned with imagery—is the mere *entertaining* of images: there are familiar and great differences in what we *do* with and through images, symbols, myths and metaphors. Think, for instance, of imagination in a worshipping, or a meditating or wondering or celebratory mode: and contrast that with imagination in exploratory, enquiring, referential and critical modes. We have to grapple with images that jar with each other and with aspects of experience. From critical interrogation of images or analogies we may emerge with more or with different and better images, better analogies, more illuminating metaphors. Indeed, we might speak of this as imagination in a rationally critical mode, or, equally well, as critical reason in an imaginative mode.

Now it is no part of our present enquiry (which is already wide-ranging enough) to argue in detail over the prospects for success or failure of a theology which sees imagination as absolutely central to theistic discourse, and also retains a transcendent reference to deity. If the pervasive presence of metaphor, imagery, myth, does not prevent rational-imaginative critique of both the *sense* and the would-be *reference* of the images, then the broad outline of my argument in this part of the paper can stand. What must be modified are the too-sharp dichotomies between imagination and judgment, all-imaginative conjecture and all-rational critique. Imagination appears on both sides of these oppositions, but it plays different roles. In the critical mode, the images do not call the tune, are never authoritative, privileged—for contemplation and to inspire: they are ancillary to an activity of thought and questioning. As Friedrich Waismann spoke of the philosopher as having to think against the current of language—'up-speech' (1956, p. 468), so we may speak of our critical mode as thinking, not just with, but maybe against the images: even though the outcome may be new images, more convincing images, perhaps less gratuitously extravagant images for the old.

Am I, in the end, saying that imagination is or is not a *cognitive* power of the mind? That sharp dichotomy of image- or symbol-set and rational appraisal would suggest that it is not. But that dichotomy has had to be considerably softened. If there is such a thing as cognition, at all, imagination will be involved in it. But cognition is not a unitary act; and imagination is one of several components. I acknowledged the (roughly Kantian) sense in which, without imagination, we cannot make the basic differentiations essential to awareness of the world and of ourselves as subjects. Its synthesizing role continues to be vital to knowledge at every level: to personal relations, where we seek to interpret disparate pieces of behaviour as manifestations of a single intelligible character; to the scientist and the historian, who seek pattern in their data. And so with religious imagination, seeking some unity in and beyond religious 'intimations', hints of transcendence, rather than leaving these as sheer anomalies, discrete individual mysteries. That too must count as a cognitive endeavour.

But not all such cognitive projects are patently successful. A critic of theism may deny that the data here can be shown to point persuasively and coherently to a single mystery, a single deity, or may deny that problems in referring the images and myths to such a deity can be overcome. I have argued, however, that to acknowledge that failure, if failure it be, can be seen as signalling a necessary dialectical development, a requirement of imaginative logic in the religious sphere. Even though it may be one that negates an 'objectified' view of the divine, it is far from negating the life of religious imagination itself.

References

Eliade, M. 1952. Eng. trans. P. Mairet 1961. *Images and Symbols* (London: Harvill Press).

Falck, C. 1989. *Myth, Truth and Literature* (Cambridge University Press).

Jaspers, K. 1932. Eng. trans. 1971. *Philosophy*, Vol. 3, *Metaphysics* (University of Chicago Press).

Kant, I. 1781. *Critique of Pure Reason*. Eng. trans. N. Kemp Smith (London: Macmillan).

Kant, I. 1790. *The Critique of Judgement*. Eng. trans. J. C. Meredith (Oxford: Clarendon Press).

Mackey, J. P. (ed.) 1986. *Religious Imagination* (Edinburgh University Press).

McIntyre, J. 1987. *Faith, Theology and Imagination* (Edinburgh: The Handsel Press).

Pascal, B. 1670. *Pensées*.

Sparshott, F. 1990. 'Imagination—the Very Idea', *Journal of Aesthetics and Art Criticism* 48:1 (Winter).

Strawson, P. F. 1974. 'Imagination and Perception', in *Freedom and Resentment and Other Essays* (London: Methuen).

Waismann, F. 1956. 'How I see Philosophy', in *Contemporary British Philosophy*, H. D. Lewis (ed.), Third Series (London: Allen and Unwin).

Williams, B. 1976. *Problems of the Self* (Cambridge University Press).

Moral Values as Religious Absolutes

JAMES P. MACKEY

I. Religious Absolutes

Those who have had the benefit of a reasonably lengthy familiarity with the philosophy of religion, and more particularly with the God question, may be so kind to a speaker long in exile from philosophy and only recently returned, as to subscribe, initially at least, to the following rather enormous generalization: meaning and truth, which to most propositions are the twin forces by which they are maintained, turn out in the case of claims about God, to be the centrifugal forces by which they disintegrate. In simpler language, the greater the amount of intelligible meaning that can be given to the idea of God, the less grounds there would appear to be for assuming let alone asserting, that God exists, at least as a being distinguishable from all the things in this empirical world which are the source of the range of meanings available to us; on the other hand, the more we insist that God exists, a being over and above the things that make up this empirical world (the more we take the proposition 'God exists' to be a true proposition in this *particular* transcendent sense, for the adjective 'transcendent' has many uses) the less the amount of commonly available meaning we appear to be able to apply to God. Or, to put this in a manner which might obviate an obvious objection to it; either everything we know is *tout ensemble,* God, and then nothing in the world that we know is distinctively divine; or else nothing in this world is God, and then nothing that we appear to be able to know is God. That same formulation will work, it should be noted, even if we substitute for 'things in the world', 'an aspect or aspects of things in the world'.

That dilemma would appear to have been present from the beginning of Western philosophy, with the pre-Socratics—and this was also in actual fact the beginning of Western theology. The dilemma emerged fairly explicitly, I would think, towards the end of the pre-Socratic era, and it is equally present, and much more explicit, in most recent philosophical writing. In fact Professor Hepburn's paper to this conference provides an example of its presence which could hardly be closer to us in time and space. The dilemma poses a far more devastating threat to religion than do the efforts of A. J. Ayer, for example, to prove all religious language nonsense, in the strictest sense of that word; for the upshot of this dilemma would be that any escape from A.

J. Ayer's assault upon the meaningfulness of religious language would only land the would-be believer in worse trouble. Dilemmas, like pincer movements in matters military, are the deadliest forms of attack upon the most determinedly held positions.

The pre-Socratics in general were in pursuit of the *phusis tōn ontōn*, but since the idea of *phusis* to them encompassed the notion of *fons et origo*, and not a little of the teleological connotation it was long to retain, they are rightly inscribed in the ranks of the theologians—in fact they are founder-members of the company of students of *theologia* even before that term was coined. They may not be inscribed in the ranks of 'mere' physicists, or even of 'mere' philosophers, as some in more recent times, which have seen the separation of philosophy from theology, and of both of them from the physical sciences, might be tempted to describe them. I can find few explicit hints of our dilemma amongst the early surviving fragments of Western philosophy/theology, even though from Zenophanes onwards those who sought the divine in the nature of things often recommend a reverent agnosticism about the nature or form of the divine.

It is towards the end of the pre-Socratic era that the dilemma as such becomes explicit. Anaxagoras it was who segregated Mind from all other things, as dichotomously as the unmixed is distinguished from the mixed, and although humans have similar mental powers, the divine Mind is very much other, over and beyond the world of things, 'infinite and self-ruling . . . alone, itself by itself'. This divine Mind effectively and rationally controls the whole of reality; 'it was Mind . . . that controlled the whole whirling movement and made it possible for it to whirl at all . . . and Mind knew all the things that were then being mixed . . . and Mind arranged all such things as were then to be' (Diels, 1935, Anaxagoras B 12). The universe is then rationally, purposefully, that is to say, teleologically controlled in its entirety. Diogenes pursues the quest for teleological planning into the very details of nature. His success could be described as progress in knowing the divine Mind. But it could also render the hypothesis of a divine mind entirely otiose. For what he was discovering was the rational plan by which this world cohered and proceeded. In other words, the rationality appeared as an attribute of the world, not necessarily, if at all, as a quality of a divine mind. Diogenes appeared to suppose that the divine Mind was responsible for the beginning of a structured world; but that supposition might also be seen to be otiose: what is clear to all is that the world so ordered *is* an existing entity, and all that is known from it of rational ordering is perceived to be within it. Hence what is known exists, as world, and if something exists over and beyond world, it appears not to be known. Thus did the very success of the first natural theology slowly but inexorably lead it on to the horns of our dilemma.

A similar, and similarly inexorable process took place in the case of the Sophists, a group to whom Socrates in his mentality and methods was closer than his subsequent admirers wished to admit. For the sophists the human *phusis* is of central significance, and as they situate religion, its origin and content, in their account of the development and structures of human nature, our dilemma re-emerges in a different, but only slightly different form. For whether religious belief is accounted for in what we today would call psychological or moral terms, the end result is the same. Democritus saw the origin of religion in visions and dreams, Critias in the *Sisyphus* extolled the prudence of having a divine, omnipresent overseer of human moral conduct, one who would combine in one person the legislative, judicial and executive functions. In both cases it is what humans perceive as products of their own psyches—spiritual presences in visions, obligatory moral rules that are not ever really ignored with impunity—that provides the known content, and an existence of something above and beyond this, acting as its putative source, would appear to be correspondingly unknown. Protagoras drew the correct conclusion from all of this, and one quite in line with our dilemma, when he said: 'I am unable to discover whether they (i.e. the gods) are or are not' (Diels, 1935, Protagoras B 4).

Kant, of course, is the main philosopher in the modern era whose *Critiques* could be seen, from one point of view, as a sustained attempt to prove the inevitability of our dilemma. The most Protestant, indeed in one sense the most Calvinist of philosophers, he resolutely confines the possibility of knowing to the empirical world, to those processes which are exhausted by the 'intuitions' to which this world gives rise, and to the 'forms' of sense perception (space and time), the 'schemata' (of cause, substance, reality and necessity) and the 'categories' (of quantity, quality, relation and modality) by which this world is known. Correspondingly, he consigns God to the status of a postulate, to faith as distinct from knowledge or, as he frequently suggests, and I think with a good deal more plausibility, to hope. I am never quite clear about Kant's idea of a postulate, but there are passages, such as the following, in which he appears to describe the very activity of postulating: 'The righteous man may say: I *will* that there be a God . . . I firmly abide by this, and will not let this faith be taken from me; for in this instance alone my interest, because I *must* not relax anything of it, inevitably determines my judgment, without regarding sophistries, however unable I may be to answer them or to oppose them *with others more plausible*' (Kant, 1956, pp. 241–2). It all sounds a little petulant to me; certainly an exercise in hoping rather than knowing. Kant was the clearest of those who saw that the ontological argument is always involved, however implicitly, in every attempt to prove God's existence; and since he saw the ontological argument as an attempt to define divinity, necessary for all who would try to prove God's existence; and since he established on the basis of his own epistemology that no proof of God's existence could possibly succeed, he illustrates better than any other single philosopher the dilemma of the believer: that what is known cannot be God, and if God exists, God (and God's existence?) cannot be known.

But if Kant was clearest on this matter, I hope I have shown that his was not the only philosophical position from which our dilemma emerged, and perhaps I can end this first section by showing from two contemporary examples that philosophers who are not particularly Kantian face us with dilemmas that exhibit very similar pairs of horns. Before he wrote *God and Philosophy* Anthony Flew went on tour with a paper entitled 'The Presumption of Stratonician Atheism'. The shrewdest and indeed the central move in that paper was as follows: he accepted the unsatisfactoriness of infinite regresses in attempts at explanations of worldly phenomena; but he insisted that any talk of an ultimate explanation could refer only to the last and most comprehensive explanation which had *de facto* been found. He suggested that such an ultimate explanation would most likely consist of some general scientific formula about worldly matter and motion; but his general contention did not stand or fall with that particular suggestion. His general contention was this: there really is no point in arguing that a particular formula or fact, the most comprehensive or final or ultimate explanation of phenomena thus far found, unless this too was patient of further explanation, would stick in our philosopical craw as a 'brute fact'; for whatever alleged ultimate we feel impelled to postulate in order to fully explain the phenomena would, by sheer dint of being declared ultimate, be patient of no further explanation and would also, in that sense, be a brute fact. At least with intra-mundane ultimates one knew what they were. With extra-mundane ultimates, postulated purely in the name of some allegedly mandatory intelligibility, or on the invocation of some principle of sufficient reason, one did not have even that advantage. The brute fact you know is, like the devil you know, miles better than the brute fact you do not know.

I believe I could take for my last example one feature of Professor Hepburn's paper to this conference. I do not wish to comment at this point on his fine critical analysis of the claims which Professor McIntyre and I have made for imagination in the treacherous business of seeking a knowledge of God. But there is a kind of refrain in Professor Hepburn's paper which I hope I may characterize, without caricature, as follows: yes, indeed, he seems to say, such claims about the nature and activities of imagination might well succeed in the sphere of religious belief, if we had something additional, some actual evidence for the actual existence of a divine object which we could then claim is in this way, however tentatively, apprehended. But, he implies, no such evidence is produced in the course of these excursuses on religious imagination. Hence we know much about imaginative ways of knowing, but we do not know if any divine object is ever known by these. What we claim to know in imagination is, then, not known of God; and assumptions about God's existence are left without the evidence which would make at least that much about God (viz. God's existence) known.

There is, then, I would say, an impressive body of philosophical evidence, both early and late, and all of it converging upon the impression of an apparently inescapable dilemma: what one can claim to know is not God; and to assert the existence of God is to assert the unknown.

Since the dilemma has been around for so long, and has often been recognized in one way or another, it would be surprising indeed, if one hundred and one efforts to evade it were not visible over the course of the history of Western philosophy. Postponement to another life of any real knowledge of God—'then I shall know even as I am known'—was one obvious option, a kind of eschatological realization of what could in this life be no more, and no less, than a rather contentless hope. But even here philosophical commentary, unwittingly perhaps, emphasizes the enduring inevitability of our dilemma, when mainline theologians like Aquinas, for example, insist that our very powers of knowing, in particular our light of reason, would have to be themselves raised supernaturally to a higher state, divinized one might say, for in their natural state they are not capable of knowing God *in se*.

Theological efforts to evade the dilemma stressed the exclusively supernatural status of the sources of knowledge of the one, true God, but when it came to the content of the knowledge thus vouchsaved, these theologians found themselves faced once more with the dilemma: in so far as the content of the revealed knowledge of God was intelligible, it would appear to have mundane reference; and so some of them were driven to saying that God revealed only that he was revealing! Such efforts, so similar in fact to the philosophical *via negativa*, turn out to be ways of impaling oneself rather precipitately on one of the horns of our dilemma; whereas further traditional suggestions from the philosophy of religion—the *via eminentiae* and the way of analogy—achieved the same effect, but much more gradually. The closer one approached distinctive divine existence, the more rapidly did one's actual knowledge decline to vanishing point.

There were always, also, what might be called epistemological attempts to evade dilemma, though they were not always conceived by their authors for this purpose. Certainly Plato appears to have felt that once he secured the distinct existence and superiority of soul over body, religious belief was safe and true; and, very generally speaking, the greater the role given to distinct and distinctive mind as source of knowledge or truth, the easier it appears to be convinced that something supra-mundane and very possibly supernatural is within reach. But of course the dilemma simply re-emerges in another form, for here also the closer one appears to approach *divine* mind or spirit or soul, the thinner does the accretion of knowledge become. It has been noted that absolute idealism and naïve realism are so close epistemologically as to be effectively indistinguishable. The describable content of knowledge appears to be roughly equivalent in both cases; the unity of the apperceptive ego, especially when that is taken on a cosmic-individual rather than an empirical-individual scale, is just as dogmatically described in one case as it might be denied in the other. But I must here defer to *The Vindication of Absolute Idealism*, by Professor Sprigge and to his analysis of the concept of the supernatural during this conference.

I do not intend to survey all of these ways in which evasion of our dilemma has been attempted, nor indeed to analyse any of them any further. Instead, for the second part of this paper, I would like to look at morality for a possible hope of evading impalement. Not, I hasten to add, at the moral experience as a staging post for a proof of God's existence; but rather at the nature of moral discourse for a clue to a possible kind of knowing which might avoid our dilemma. For I hope it is by now accepted that most of what philosophy has counted as ways of knowing has, in the case of God, left knowledge behind as it appeared to approach actual existence, and correspondingly, the more confident it was of the content of knowledge the more it appeared to be talking of quite mundane existence.

Now moral discourse appears at first sight to offer an intriguing similarity. The old 'no ought from an is' adage, at least in some of its less critical usages, seemed to suggest that knowledge of empirical reality could not be turned into knowledge of moral duty or moral values—yet no one as far as I know has ever doubted that we do and must know a great deal about the concrete content of moral values, and that these values contain some reference to existence, over and above the merely notional existence which the mere fact of our thinking about them confers. The nature of this reference and the kind of existence it reveals are, of course, highly controversial matters. On the other hand, the fact that in the Western tradition, if only as a consequence of the permanent dominance of Plato, terms for prominent moral values, such as goodness and truth, have been used interchangeably for God, holds out some hope that a careful analysis of moral values, of their meaning and truth, and more particularly of the coincidence of their known content and of their reference to existence, might throw some light upon the dilemma about the meaning and existence of God.

II. Moral Values

Any approach to the meaning and existence of moral value must begin with a fresh critical look at the adage 'no ought from an is'. The acceptable meaning of the adage, I think, is this: no factual and value-free description of a state of affairs that happens to be so can ever of itself entail that that state of affairs ought to be (so). But the adage cannot be turned about and driven in the opposite direction; it cannot be taken to mean that in describing a moral value I am never describing an actual state of affairs; for moral values, in so far as they are realized in the world—and all moral values either are or have been to some degree realized in the world—describe states of affairs actually brought about or maintained in the world by actual moral agents. They may, in fact they commonly will, describe more than that, but they always describe at least that. I do not wish to pause at the moment to consider what on the continent might be described as the epistemological preference for praxis (see Marx's *Theses on Feuerbach*) or what in this country is called the practicalist thesis (by Gilbert Ryle, for example), albeit this thesis might well mean that moral knowing is in fact the commonest form of the acquisition of knowledge, and not, as is more often thought to be the case, that more speculative, inductive and deductive processes provide the most general means of the acquisition of knowledge, while moral intuition, perception and conviction fall into a distinctive, and distinctively problematic category of knowing. Such a thesis, if it were to be successfully maintained, might prove to be of inestimable value to the analysis which I attempt in this part of the paper, but I wish to pause here only to note that moral knowledge can be seen, despite the first uncritical impact which the 'no ought from an is' adage may have upon us, to combine in itself just those elements of empirical content and distinctively moral reference, while the corresponding elements in the case of God-talk, empirical content and distinctively divine reference, seemed destined to fall apart.

Perhaps, however, this combination continues to be a feature of moral values and of our admittedly vacillating apprehension of them because, unlike allegedly religious ideas, they never threaten to take us beyond the empirical world. They often, perhaps always, refer to better states of this world than we have yet seen, but they never of themselves refer to a world other than this, or to a being over and beyond this world. The only way to answer the query implicit in that suggestion is to analyse more closely the reference to reality already said to accompany a moral value, and to ask in particular if adjectives like 'absolute' or 'ultimate' have any legitimate place in moral discourse.

Is there, then, a distinctive kind of reference to existence or reality which might be thought to characterize moral discourse and which

James P. Mackey

might indicate some way of avoiding the 'meaning or truth' dilemma which seems to affect religious language? One might approach an answer to this question by attending first to what is commonly called the sense of obligation which accompanies all actual moral states and acts, and which is represented in moral discourse by the term 'ought'. There is some reason to think that far too much time and energy in moral philosophy is devoted to the analysis and defence of freedom, on the supposition, no doubt correct in its way, that if the moral agent did not enjoy some measure of freedom, moral discourse would remain bereft of the ability to denote. But it may well be that this elusive quality of freedom is best approached through an analysis of what is called the sense of obligation, often, comparatively speaking, neglected. For what we call the sense of obligation is, as its name implies, something of which we have an empirical awareness, in a manner perhaps in which there is no corresponding awareness of freedom. When we say 'I feel free' or words to that effect, we seldom refer to anything more than a permission given or, more generally, to the felt absence of external compulsion; and moral freedom must mean much more than that. What is called the sense of obligation, on the other hand, is surely experienced as a kind of compulsion—better, perhaps, impulsion—which is compatible with, and indeed may require a certain degree of creativity. The image which, in the history of Western philosophy, would then come closest to elucidating the nature of moral obligation would be the Platonic *eros*. The element of creativity, in any case, provides the next clue to the distinctive kind of reference to existence which moral discourse may be thought to entail.

The creative impulse, the obligation of which I am aware, alerts me to the fact that the state of affairs in which I find myself or the behaviour in which I am currently engaged, is one that not only is but ought to be (or, of course, ought not to be, as the case may be). There is thus at this first stage a high 'is' content in every 'ought'. I am a participant in this conference and I am reading this paper to you, and given my profession and the nature of philosophical inquiry, so I ought to be. There are, of course, more basic, more serious examples. I live with a woman and we have two children that we feed, clothe, shelter and try in vain to educate. I live on this lean earth and try to take my livelihood from it without destroying the very prospect of my future livelihood and that of others. So many and such general states of affairs which exist; such a confluence of behaviour patterns which at this moment actually occur; all can be empirically described because they exist or occur, and it can equally be simultaneously said of them that they ought to be. Now since the impulsion or obligation is itself an empirical datum, in both aspects at once these states of affairs and behaviour patterns are patient of empirical description. Ought and is coincide at least in the opening stages of any analysis of morality which begins with the so-called sense of obligation.

The first moment at which the ought can be felt to outstrip the is, in the sense of is which connotes contemporaneous actuality, is the moment of realization that a state of affairs or an activity which is and ought to be, should perhaps be maintained for the future because it ought to be. It is at this moment that the element of creativity shows itself more clearly. It is not at this moment that the element of creativity is first present: the human species is the least specialized in all this world. Other species whose organs, attributes and various appendages are specialized can be confined in their contributions each to a particular ecological niche or stage in the evolutionary process. But the unspecialized can meddle in everything at every time and place, and does not have its contributory work dictated, as it were, by the ecological structures from which it takes its organic form. So the human creative contribution is already present in all the states of affairs and behaviour patterns already described as actual. It is just that it is not likely to be noticed before the issue of maintenance explicitly arises. For the question 'to maintain or not to maintain?' introduces the realm of possibility, of that which might or might not be, and of that which could be other than it is. To an agent so unspecialized by the ecological niche in which it finds itself—and whose brain is as unspecialized in that evolutionary sense of the word, at least in some parts of it, as any other organ or limb—to be faced with future possibilities is in effect to be alerted to the fact that present and actual forms of co-operation in the world were not predetermined either.

The moment one faces possibility, that is the moment in which the specific nature of human *eros* or impulsion comes home to one: an impulsion that is aligned to real creativity and hence contains the element of freedom. Angst occurs at this moment also, according to Kierkegaard and his existentialist followers. But what is of more interest to us in the course of this analysis is that the reference to reality, the 'ought to be' factor explicitly emerges at this moment also as perhaps the human person's first real intimation of transcendence—and I take transcendence here in the simplest sense, to indicate the possibility of going beyond the *status quo*. Before analysing the nature and possible extent of this transcendence; before asking in particular if it can ever attract the adjectives 'absolute' or 'ultimate', and if so, in what sense, it is best to deal briefly with some more pressing problems.

First, one may ask, in the first moment of realized transcendence (in the simplest sense of that word just now adopted) what is the nature of the reference to existence which moral discourse now reveals: it is not now reference to what ought to be as it currently is; it is reference to what ought to be the same or other in the future, but just as surely ought to be. Can this more obviously transcendent reference to reality in moral discourse be called a claim to exist, or even more strongly in quasi-juridical terms, a 'right to be'? If so, it must be reasonable to ask: who or what is the source of this obligation that states of affairs continue to be or to change, or to change the metaphor, against whom or what can claims or rights to be, be urged?

As to source, I think obligation is as original a feature of the experience of being a human person as is knowing. In short, just as one cannot go behind one's actual knowing to ask how it is possible to know, where the power to know comes from, any more than one could move out of one's body in order to answer a question as to how one got into it, neither can one get behind the sense of obligation to see its source. It simply is a feature of the nature and existence of what are called persons, and through their part in reality it makes the world moral. One may liken it to other forms of apparent impulsion, treat it as a special case of some general, cosmic *eros*, but none of this will reduce its status as an original, originating fact (in Flew's argument, a brute fact) of personal existence. Indeed if the practicalist thesis is true, then the two cases I have just taken, the origin of knowledge and the origin of obligation-awareness or sense of obligation, are not really two but one. If 'knowing that' always succeeds 'knowing how (to do something)', then it is the originating creative and, to that extent, free impulse to do, to maintain, or to make to come about, that is the original and inexplicable origin of all knowledge. (Two incidental corollaries might be noted at this point. First the quest for knowledge is as moral, or as immoral an enterprise, in general and in particulars, as any other enterprise. Acknowledgment of this would surely relieve us all of the quizzical sight of scientists claiming that the alleged advance of knowledge in which they are engaged is morally neutral, if not always good, and only the practical use of this knowledge or its application need be the subject of moral criteria. Second, since the 'ought' is an original and originating datum *sui generis*, it cannot be said to be derived from an 'is'; the 'ought' will always, of course, have empirical content, for what ought to be will be either actual or entail a claim or a right to be in the future. So in a sense it is the 'is' which is derived from the 'ought' as far as moral agents in the world are concerned, and as far as a world acted upon by moral agents is concerned.)

Against whom or upon whom, then, is this 'ought', this right, urged? The answer can only be: upon persons, for the simple reason that the sense of obligation is one of the defining factors, if not the principal defining factor in the notion of a person. So there is a claim upon persons that some states of affairs be maintained or brought about; with the proviso perhaps that none of these states of affairs be existential or metaphysical impossibilities. There is as yet no intention of suggesting the existence of some transcendental Person (with capital P) against whom all claims to exist or to come about might be validly raised. We do, each one, sense an obligation which we discover we have in common with others, which therefore seeks to realize itself in values that relate us to, or that we must have in common with others, and yet no one of us feels himself or herself the sole source of this common and distinctive *eros*, or solely responsible for its communal goals. But may we then leap to some holistic view in which we must appear to be instruments or aspects of an ultimate or absolute Moral Being whose very nature it is (as the hyperessential Good of Plato) to diffuse itself creatively?

Consider for a moment the term 'absolute'. The term *is* used in moral discourse; one does hear much talk of absolute values. That in itself, of course does not guarantee that such talk makes any sense in moral discourse; nonsense is everywhere. Are moral values ever absolute, and if so, in what sense?

Moral values are states of affairs, obligatory and actual or obligatory and possible, which exist, are maintained, or may come to be under the free, creative impulse recognized as our sense of obligation. The term 'values' usually refers to rather general and abstract notions such as justice, truthfulness, and most abstract of all, goodness itself, as aspects of states of affairs. Because of their huge generality, such terms require more specific formulation if moral agents are to know in any useful manner how to live as they ought, and this more specific formulation can be provided in a number of ways: by concrete picturing or story-telling or, more usually though less popularly perhaps, by the construction of codes. All of these latter are then ways of specifying moral values, and all of this is in mind when the question is asked about absolute status for any of it.

Just as the origin of moral values lies in itself—that is to say, states of affairs existing, maintained or envisaged under the creative impulse we know in our sense of obligation, originate in the creative impulse realized in the states of affairs so existing, maintained or envisaged—in the same way as the origin of knowledge, of which moral behaviour may be the principal part, lies in knowing; so the goal or purpose of morality lies also in itself. Goodness is its own reward; to be good (well-being) has no ulterior purpose. In that sense morality in general is absolute, in the sense that the totality or ensemble of moral states of affairs, as described above, derives from nothing anterior to itself and has no purpose, goal or end beyond itself: morality in general is not relative to any source or any goal. That is sometimes represented as the autonomy of ethics: but autonomy here is not the negation of theonomy, any more than it is the assertion, without more ado, of theonomy. It is, or at least it is meant to be, the result of a simple analysis of the structure of moral values which takes its starting point and direction from an analysis of the sense of obligation.

Such a sense of absoluteness, of course, refers only to the huge and developing totality, morality itself, or morality in general. But concrete moral values, particular states of affairs which ought to be—can any of these ever be described as absolute? It would seem difficult to give an affirmative answer to this question. What it is good for me to be doing, to keep on doing, to begin doing, would seem to be always relative to the actual, and usually quite physically circumscribed possibilities of a particular place and time. [I am not raising here the question as to how I know that x is good for me. Moral philosophy often concentrates, and sometimes exclusively, on that question. I assume the answer has much to do with trial and error, or at least with experience of life enhancement and diminishment and analogies that can be drawn from these.] But what of something such as 'life itself', something not quite as abstract and general as morality or goodness, yet not at all as circumscribed as any of the myriad and more concrete moral values which may, as the saying goes, make life worth living?

The claim that life itself is an absolute value is usually restricted, of course, to human or rational life which is itself, and then makes its world, a series of states of truly moral affairs, and the claim is sometimes represented in codified form by a precept which forbids the deliberate taking of human life under any circumstances whatever. There is, I think, something to be said for this: indeed it may represent a derivative from, or even a version of the case for the absoluteness of morality in the sense already explained. There would presumably be problems about the restriction to human life, or even to rational life, but these too would coincide with problems in the assertion that morality, the total thing, was itself underived from and undirected to anything in existence other than itself. [In terms of this empirical world, morality would either originate in a quantum leap coincident with the *mise en scène* of humanity or of any other life form which could be shown to be capable of moral agency, or some early Stoic universe would obtain in which *logos* was immanent and the whole of reality thereby made moral from the outset.] But I prefer to concentrate in this short paper on a different kind of problem affecting this contention— the contention that rational life is an absolute value—one which derives from the nature and inevitability of death.

It might appear that because of its utter banality and unconditional inevitability, death and dying should be withdrawn from the category of states of affairs which are or embody moral values, on the grounds that it is not something we do; it is purely and simply something which happens to us. One simply has to state the position to see that *it* will not do. Apart from obvious cases in which we deliberately take the lives of others, in war for example, or take or give our own lives, there is a more general case to be made for saying that dying is something that all of us do all the time. And that not merely in the sense of preparing for our biological death, whenever or however that may come about; but in the much, much larger sense that a disconcerting number of ways in which we appear to describe living could, with but little alteration, or none at all, be taken as descriptions of dying. 'We live out our years on this lean earth', for example, or talk of any of life's activities which, while they build us up, simultaneously wear us down.

All of this means that the 'is' which ought to be, the continuous creation, the states of affairs constantly coming to be under the influence of the underived 'ought', all are maintained and advance in face of an incursive nothingness with which they are, however, in deep complicity. But perhaps we may say that it is nevertheless to the 'is' that the ought is directed, not to the non-existence which is apparently an unavoidable by-product of activity for each individual agent. It is existence that ought to be, and existence that ought to be is moral existence, a state of affairs or states of affairs that result from obligatory impulsion, erotic obligation (in the comprehensive Platonic sense of *eros*, in which it is not exclusive of *agape*). This, I think, yields two different kinds of conclusion.

First, that individuals may never deliberately bring about the death of moral agents, for it is on such that the maintenance of moral reality depends; and it cannot be morally right to destroy morality, or any part of it; and they may only take such action or stance as could threaten their own lives, in order to maintain and advance moral living for the other. Of course, that precept always comes too late for all of us, for everything that exists. The very first fragment of Western Philosophy now extant, from Anaximander, talks about *ta onta* 'the things that are' having to pay the penalty and make atonement to one another for their injustice (Jaeger 1947, pp. 34–5)—a truly extraordinary piece which seems to anticipate the long and tortuous and most recent metaphysical analyses of Levinas, and all his talk about the way in which each existing thing is already laid claim upon by others and indeed accused by these others of taking place (literally). We all of us exist at the expense of others and that is an original fault for which we can only ask them for forgiveness and make some recompense in being responsible for and towards them. So the offence merely confirms the rule, the obligation inscribed in the heart of reality to wager for life and life more abundant, for being without end.

Second, this brings us back to the contention that moral being itself is the absolute, the only absolute to which we can give a continuously self-transcending content of meaning. Is this God, then? The analysis of ideas so far conducted would not give grounds for a simple answer to that question; all that can be said reasonably is that a defensible and meaningful use of the term 'God' could be found at some point of the continuation of this analysis, and in particular, on further analysis of the idea of *eros*. If it is accepted that morality is best understood by analysis of our experience of obligation: if obligation is a kind of impulse/attraction which in creative agents entails measures of freedom; if the impulse/attraction named *eros* can be seen, first, to arise *within* us, certainly, but not noticeably *from* us and, second, to name a kind of pervasive force in the universe out with our human agency; if we are enticed by our awareness of this pervasive force to seek to know how much of its presence is indicative of, or how long it has been characteristic of that kind of free creative being we call 'personal'; or, to put the matter in a slightly different way, if we can attempt to estimate, on the evidence we have, how much of the reality we are aware of, past or present, can be accounted moral states of affairs because of the presence of this *eros* operating through and as moral agency; *then*, since we can already argue that such moral reality alone attracts the adjective 'absolute', in the meaning already outlined, we should probably be in the presence of the only philosophical case we can make out for belief in God. We should then, of course, have landed back in our laps all the usual problems about gods and their relationships to the world. Is Eros god, and the ensemble of states of affairs which make up the totality of the empirical world her 'body'? And what of the relationship of each of us individual persons to this god? Are we, each and all rational agents, special 'incarnations' of this universal and original divine force, or power, or as religious folk would prefer to say, 'spirit'? Or participants in it, or instruments of it, or no more perhaps than indications of its presence and of its forever passing on its eternal way?

It has not been the task of this paper to raise, much less to answer these questions, or even to take us very far along the path on which they eventually arise. We do not even seem to know, and perhaps we cannot know for sure, whether a moral universe actually existed before the origin of our species, and we have no way of knowing whether the moral existence for which we wager with every breath and every move will actually survive the death of our species, will reach a stage in which the non-existence (death, to us) which is now engaged in every effort to continue to exist will no longer threaten some future moral reality; but it is precisely for that that our free creative impulse, our morality, itself constrains us to wager and for that, therefore, it enables us to hope. Yet we have no way of knowing if our individual persons can hope for any permanent place in that unthreatened moral universe which christians call eternal life. All that we can be sure is our part in wagering for it while we live and in the enjoyment of such partial and temporary realizations of it for as long as we may. The simple, and sole contribution which this paper can hope to make is this: to show how one can set out on this path in such a way that meaning and truth do not part company as one proceeds; alternatively stated, to seek to depict the process of knowing reality so that issues of meaning and truth can still be combined even on the approach to, or of 'the absolute'.

III. Conclusion

I realize that I have traversed great distances at such a speed that, had I been in control of a mechanically propelled vehicle, I would be subject to a charge of reckless driving. To summarize at this point might seem to compound my guilt; but I have little option, and if I have caused or do now cause some serious philosophical accident, I will at least remain at the scene and accept full responsibility.

I would seem to have envisaged an epistemology in which knowing how is indeed the prior part of knowing that; and since knowing how involves doing, at least in the form of practical apprenticeship, it involves already what we call morality—that is, action under a conscious sense of obligation which still leaves us, as moralists say, free. Hence an adequate account of that elusive sense of obligation in action, hitherto confined to some obscure section of moral philosophy, would in fact serve as an account of how we know our world, and indeed all of reality. My interest in this is to reveal, if possible, a different relationship between knowing and being, meaning and truth, from that which seems to disintegrate on some traditional understandings of these whenever divinity or the absolute comes into view, or tries to come into view.

Most theories of knowledge, and certainly those on which meaning and truth disintegrated on the approach of the divine, operated upon a distinction between the knowing subject and known object, whether that object be other things or other minds, and the problem then was to find another (thing or mind) which could be called divine, amongst all the things and minds one knew or might come to know; such theories of knowledge, in other terms, operated upon the visual or contemplative model. But if morality, activity under the impulse of the sense of obligation, is the source and heart of all knowing, then knowledge is participatory rather than contemplative, an awareness (at some times more explicit than at other times, but always capable of reflective deployment and precision) of participation in a cosmic process driven by an originating, but not originated eros-type impulse, and unlimited, since it is not a means to any further end. The relationship of such knowledge to existence is quite different from that of knowledge conceived on the visual or contemplative model; for it is now the self-awareness of a part or instrumental agent in a reality-process that vastly transcends it, but of which it is thus both aware and knowledgeable. In one sense Kant's idea of knowledge was also this contemplative idea; it was just that, in the *Critiques*, he defined it impossible by reason of the intervening forms, schemata and categories; and it took Hegel to remind us that Kant could not have known all of *that* unless he, and we, had some deeper and more direct awareness of spirit moving in the world. Known in this way, then, by us humans, who are parts or instrumental agents of it, reality actually appears, however dimly, as this driven thing, without derivation and without goal or end beyond itself. In short, in moral knowing, knowledge and being do not fall apart as being begins to appear absolute; and yet what is known in this way can scarcely be described as a thing over and above other things, nor would it appear adequate to refer to it merely as an aspect of a thing or things—it is too much like a source of real states of affairs, or at least of their maintenance, renewal or change.

None of this, of course, is new. When the Platonists described the process of creation itself as *bonum diffusivum sui*, the hint about the intrinsically moral nature of being and of knowledge was given. Indeed when Hesiod's *Theogony* named Eros amongst the oldest of the gods, the clue to the nature of the divine and of our participatory familiarity with it was already in place.

References

Diels, H. 1935. *Die Fragmente der Vorsokratiker*, W. Kranz (ed.), (Berlin).
Kant, I. 1957. *Critique Of Practical Reason* (New York: Liberal Arts Press).
Jaeger, W. 1947. *The Theology of the Early Greek Philosophers* (Oxford).

Revealing the Scapegoat Mechanism: Christianity after Girard

FERGUS KERR

I

The philosophy of religion, as commonly understood by Christians in both the Catholic and Reformed traditions, whether they think it a worthwhile enterprise or not, begins with arguments for the existence of a deity, proceeds to show that this deity is necessarily unique, eternal, and suchlike, and leaves it to reflection on divine revelation to consider whether this deity might be properly designated as 'three persons in one nature'. Much later, after discussing the metaphysical implications of the incarnation of the second person of the triune godhead, one would arrive at theories about the death of Jesus Christ as putatively redemptive, and describable as sacrificial, atoning and the like.

Some good work has been done recently on how, when and why this paradigm established itself in Christianity. Plainly, its origins lie in Greek philosophy, as L. P. Gerson (1990) has recently demonstrated with great thoroughness. Subsequently, at the Enlightenment, philosophers started to look for natural explanations for the existence of religion. The supernatural claims of Christianity rapidly became a matter of secondary interest as all the intellectual energy went into discussing the rational foundations of theistic belief, as Buckley (1987), Preus (1987) and Byrne (1989), among others, have recently shown.

Suppose, however, that, instead of beginning from what is logically antecedent and perhaps even extraneous to the Christian religion, one were to focus straightaway on what might seem to an outsider the most arcane and esoteric, if not even the most implausible and unpalatable, of all Christian doctrines—that the crucifixion of Jesus of Nazareth is to be regarded as a 'sacrifice' which 'reveals' our 'sin' and offers 'redemption': what sense might we derive from such a proposal?

What René Girard offers, after a lifetime as a literary critic, is a reading of the Bible which makes of it, in effect, the most 'revelatory' corpus of literature in Western culture—indeed, the Bible becomes the story which reveals the 'sin' upon which our culture—any culture—is founded. This seems, at least, an interestingly different form of 'natural theology'. Instead of arguing over the hypothesis that there is a deity, that is to say, we might do better to uncover the presence of religion in any and every human society and assess its effects.

II

Girard's finest work is undoubtedly his recent book on Shakespeare (1991): the fruit of a lifetime's reflection. For the purposes of this paper, however, the essential text is *Things Hidden since the Foundation of the World* (1987), a revision as well as a translation of the original French version (1978). Girard's theory has been discussed since the early 1970s, particularly in French and German Catholic theological journals. By the 1980s the discussion had spread into North American journals, and several of these papers, particularly those by Galvin (1982), North (1985) and Wallace (1989), offer the best access to Girard's work and show how challenging it is proving to be in Christian theology. Girard's thesis may, of course, turn out to be unsustainable, but there is no doubt that it has renewed theological interest in the significance of the death of Christ.

This is not to say that the significance of the crucifixion of Christ is a new topic in Christian theology—far from it: it is plainly the crucial event in the Christian scriptures. Nor is it new to relate the central event in the Christian texts to analogous events in other religions and indeed in pagan mythologies. Since the emergence in the nineteenth century of social anthropology and the difference that has made to the theory of religion (though not to the philosophy of religion), the phenomenon of sacrificial death has received a good deal of attention. Frances Young (1975) brings her scholarly studies of sacrifice in early Christian literature to bear, in a very fruitful and often moving way, on patterns of ostracism and victimization which are familiar in everyday life—all quite independently of Girard's work. Again, more recently, Godfrey Ashby (1988), also without reference to Girard, speaks of 'much work' having been done 'to rehabilitate sacrifice'—which he regards as 'perhaps the most basic and widespread phenomenon in religious experience' (certainly a controversial claim). Finally, Stephen Sykes (1991) has edited a very substantial collection of essays, the product of a decade's work by scholars at the University of Durham—again quite independently, although Sykes notes in his introduction that 'the controversial and wide ranging work of René Girard . . . will ensure the continued currency of sacrifice as a theological issue in the next decades'.

Girard's project is thus only one, if no doubt the most ambitious, in a variety of recent attempts to re-focus Christian theology on sacrifice. What distinguishes Girard, however, is the claim that the Bible is the privileged locus of a liberating 'revelation' of the origins of culture in religion. His philosophy of religion is simultaneously a liberation theology.

III

Girard's story goes somewhat as follows. He starts with the phenomenon of imitation, with which we are all familiar. As Aristotle noted (*Poetics* 1448b 5–10), imitation is second nature to human beings from childhood, one of our advantages over other animals being that we are the most imitative creatures in the world. It is by imitation that we first learn. Indeed, but for this capacity to copy and mimick, we should never learn anything. Mastery of language, for example, depends on our aptitude for imitation. So far the story seems simple enough.

Girard insists that desire is as radically dependent on imitation as anything else. We learn to desire by identifying with someone else's desires. He or she is our model. As infants, for example, we enter the family circle by modelling our desires on those of our elders and siblings. We do not desire spontaneously, autonomously, as it were ex nihilo—we copy other people's desires. Indeed, we find it extremely difficult not to imitate other people in their behaviour, use of language, repertoire of gestures, and so on. It is a very simple claim. From Kiev to Kansas City, for example, every young person seems to want jeans and trainers. Our desires are generated by models. The advertising industry depends on this. Much more profoundly and pervasively, in such practices as apprenticeship and discipleship, imitation seems constitutive of anything that we should recognize as human life.

So far so good. The next step is more controversial. At the very moment when we start desiring what our model desires (or already possesses) we and our model become rivals, wanting possession of the same thing—the same toy, the same piece of cake, the same man, the same woman, the same job. Our model becomes an obstacle, even our enemy, whenever we desire something scarce. Once again it is a very simple claim which Girard is making. Imitation easily generates rivalry, desire which is mimetic in structure leads to conflict. Indeed, since we are educated into desiring by imitating, and thus want what everyone else wants (or already has), love, money or whatever, conflict in any human group is almost inevitable. Jealousy, theft, war—in one form or another, at one level or another—violating other people's territories, rights, bodies, or whatever, is all but inescapable. Violence is built into human life.

The story becomes increasingly controversial. According to Girard, we regularly deal with such conflict in the group by fixing on some one, some minority, as the alleged source of all the tension and expel him, her or them. Of course making someone a scapegoat need not be a conscious decision. On the contrary, it will be all the more effective if the process remains hidden. Perhaps a community needs an enemy to confirm its sense of identity. The 'black sheep' about whom the family complains all the time (or whose name is never mentioned) effectively keeps the family together. The best way of dealing with violence within a group often is to deflect it towards one of the participants. Conflict-ridden societies easily round on some innocuous subgroup and victi-mize them. Hitler's Germany rounded on the Jewish community, Idi Amin's Uganda expelled the Asians, and so on. The violence of scapegoating someone re-establishes peace and harmony in the group, or appears to do so, at least for a time. At least that is the intention. Groups need, periodically, to sacrifice a victim to keep the peace. This is the violence which maintains the illusion of a conflict-free society.

Finally, and most controversially of all, religion (according to Girard) is rooted in the conflict internal to any society. Peace has to be restored and any community will resort to familiar practices of victim-ization and ostracism. But it is the function of religion to maintain peace by institutionalizing the sacrifice of a victim. The most original rite which creates a relatively conflict-free society is the blood sacrifice. Religion indeed has natural causes, as has been argued since Bodin, Herbert of Cherbury, Vico and Hume (Preus, 1987). Far from dis-crediting religion once and for all, however, as modern atheists and humanists would think, the revelation of the truth about its natural origins shows rather that religion is an essential dimension of all human life.

There is thus no point in arguing over whether there are good reasons for going in for religious behaviour and belief. If Girard is right, religion is always with us—if by religion we understand the sacrifices which repeatedly ensure the peace of this or that society. What is distinctive about Christianity, or rather about the Judaeo-Christian scriptures (as Girard prefers to say), is that, in this body of literature, and above all in the story of the killing of Jesus, the phenomenon of scapegoating is revealed for what it is: an act of collective violence which only secures temporary respite from conflict and from repetition of which we need somehow to be 'saved'. As Caiaphas the high priest says to the chief priests and Pharisees gathered in council at an extremely important point in the story of Jesus: 'It is expedient that one man die rather than the people perish' (John 11: 49–50). The Bible, so Girard seeks to show us, is a story—indeed the story—which reveals the hidden dynamic of groups as the 'sin' it is and thus opens the way to that kind of collective self-knowledge which would free us from the compulsion to practise forms of scapegoating. The sacrifice of Christ, to use traditional language, is the sacrifice which, by disclosing the possibility of forgiveness, shows the needlessness of sacrificing others to save oneself.

IV

The interest which social anthropologists have taken in the phenomenon of scapegoating, as Girard notes (1987, 132), goes back at least to Frazer. In successive editions of *The Golden Bough* (first published in 1890), he accumulated evidence for the existence, as he thought, of an 'institution': the 'public scapegoat' employed for the occasional or periodic expulsion of evils from the body politic, typified by the priest-king slain by his successor in the sacred grove of the goddess Diana at Nemi.

To head off one obvious objection from biblical scholars, it may be noted at once that, like Frazer himself, Girard does not tie the notion of the scapegoat to the ceremony which is prescribed in the Book of Leviticus (chapter 16). He has always been aware that there are *two* goats in the story of that ceremony and that the one which is allowed to escape is the one which is *not* sacrificed in the ritual. Indeed, at his most rigorously theoretical, Girard avoids the phrase 'scapegoat' (*bouc émissaire*), preferring to speak of 'victimage mechanism' (*mechanisme victimaire*). Whatever its provenance, the notion of scapegoat, in most modern western languages, ordinarily means an innocent victim who is sacrificed as a way of dispelling some supposed evil in the group and restoring harmony. In fact Girard accepts that the word 'scapegoat' refers both to a ritual institution discoverable in many different religious traditions and also to a spontaneous psycho-sociological mechanism, a cathartic crisis, observable as a regular phenomenon in the life of many different human groups. Dictionaries would no doubt have us believe that the latter is a figurative use of the word, Girard says (1987, p. 133), so that the proper use would be in connection with religious ritual. His suggestion is, on the contrary, that societies which suppose themselves to be post-religious remain bound to the cycle of the victimage mechanism and are just as implicated as any savage tribe in 'sacrificing' some for the good of the others.

Girard is well aware that his proposal builds on the work of Durkheim. He takes for granted Durkheim's rejection of the liberal-individualist conception of society—the idea that the individual and his self-interest (and I mean his!) comprise the basic unit of society and that community is a derivative and even artificial entity. The social contract view of the origins of society needs to yield to something like Aristotle's view (e.g. *Politics* 1253a), according to which the political

Fergus Kerr

community is 'a creation of nature' because human beings are 'naturally political animals'. For Girard, as for Durkheim, society is prior to the individual. The social can never be reduced to the psychological. Then, most notably, for Durkheim, religion is best understood not by refer- ring to the supernatural (revelatory interventions from another world) or to the mystical (revelatory outpourings from the individual's soul), but in terms of the internal dynamics of the human group. Religious rituals are understood best as symbolic expressions of social bonds, religion is essentially an aspect of society.

Girard has criticisms to make of Durkheim—'Durkheim, for example, made the opposition between the sacred and the profane too absolute' (1987, p. 43) in certain of his discussions. But he was the first, according to Girard (1987, p. 63), seriously to oppose the dismissal of religion which we in the West have inherited from Voltaire and the Enlightenment—*cet escamotage sceptique du religieux*—according to which religion is a 'conspiracy of priests to take advantage of natural institutions' (1987, p. 63). This remains the prevalent assumption among educated people, so Girard says. Religion would thus be a secondary phenomenon, parasitical upon already existing social institu- tions. For Voltaire, religion was essentially a trick played by a minority upon the rest of society: priestcraft as exploitation. 'The only other plausible theory', Girard argues (1987, p. 70) is that of Durkheim: 'religion must be the origin of all institutions'. Far from being posterior to and parasitical upon social institutions, that is to say, religion is coeval with them and even foundational. Girard accepts that his work continues and completes Durkheim's project—thanks however to his discovery of the victimage mechanism (1987, p. 70).

Girard even refers to Durkheim's intuition of the identity of the social and religious domains—'which means, ultimately, the chrono- logical precedence of religious expression over any sociological con- ception'—as 'the greatest anthropological intuition of our time' (1987, p. 82). Educated people may not yet have realized this, but the time has come to abandon any view of religion which regards it as something superimposed upon supposedly more fundamental realities. Death, for example, is not to be regarded as something natural to which religious significance, rites, feelings and so forth may or may not be added; death may, and on Voltairean lines no doubt should, be desacralized, but, according to Girard, it has always already been 'sacred'. If there is a problem it is to get the sacred out of basic socio-psychological practices and institutions, not to find reasons for dressing them in religious garb.

Indeed Girard goes on, somewhat dramatically, to endorse the view that funeral rites may be at the foundation of any and every human society: 'There is no culture without a tomb and no tomb without a culture' (1987, p. 83). Graveyards and grieving, burial ceremonies and

166

mourning rituals, may thus even be constitutive of civilization—something like that.

For Girard, then, following Durkheim, religion always exists, however disguised and transformed, at the foundations of every society. It is not something which calls for justification as if people should not go in for it unless they have established its foundations at the bar of reason. On the contrary, by expressing aspects of social reality and meeting certain socio-psychological needs, it is already at the foundations of all human society. There is no culture without religion, no society without sacrifice.

Durkheim, while not unaware of the reality of social conflict, certainly emphasized the socially unifying function of religion. But violence within society is of far greater significance for Girard. One might even say that the difference between Durkheim and Girard lies in the latter's focus on how a society deals with internal conflict. Girard's discovery of the victimage mechanism makes all the difference.

One objection may be mentioned briefly. From a Marxist point of view, Girard (like Durkheim for that matter) tends to conceive conflict in terms of tension between the individual or minority and the whole collectivity, rather than between interest groups in a system of domination and subordination or exploitation. Durkheim seems nowhere to envisage the possibility that religious beliefs might be regarded as ideologies, legitimating and obfuscating the power of one group over another. For Girard, no doubt, such a claim would only be another version of Voltaire's conception of religion as priestcraft, much more sophisticated no doubt but only further from the truth. Religion would be an understandable illusion rather than a reprehensible mischief but it would remain something superstructural. If objections of Voltairean and Marxist provenance may be set aside, then, in this summary fashion, Girard's development of Durkheim's theory of religion may continue to warrant our interest. To deal with religion by taking it to be imposed by trickery upon simple people by a self-perpetuating elite may well be a tempting thought, for church-going people sometimes as well as those who are deeply opposed to religion, but it is hard not to think that religion reaches far more deeply into human life than the priestly conspiracy theory suggests.

V

But Girard's theory of religion did not emerge so much from sociology as from literary criticism. He is, after all, a professor of literature. His first book, which appeared in French in 1961 and in English in 1965 as *Deceit, Desire and the Novel: Self and Other in Literary Structure*, is a set of studies of novels by Cervantes, Stendhal, Dostoevsky, Flaubert and Proust. In the case of Flaubert's *Madame Bovary*, for example, Girard insists that what Emma desires is a Great Passion—but this desire gets into this farmer's daughter's head from the stories she voraciously consumes on the sly during her convent schooldays. Emma's desire, as Girard puts it, is mediated by the desires of others, her models, just as Don Quixote's dream of knight errantry originates in his reading of medieval romances. The protagonist in *Crime and Punishment* kills the old pawnbroker not for her money or out of hatred for her but because of his desire to be a certain sort of 'hero', proving that he is beyond conventional norms of good and evil. Stendhal's hero in *Le Rouge et le Noir* tries to seduce the two women in pursuit of an ideal of conquest, an ideal born of his admiration for Napoleon. Proust's narrator wants to see a play, go to a party, possess a woman, write a book, and so on, always because some quasi-magical prestige is attached to these things by somebody else. Girard's theory of the imitative structure of desire springs from his reading of these novels.

Girard gradually expanded his inquiries to other literary traditions and then into social anthropology and cultural studies generally. By the time that *Violence and the Sacred* appeared (French in 1972, English 1977), bringing together studies of Sophocles, Shakespeare, Plato, Lévi-Strauss and Freud, Girard had worked out his theory of the origins of culture in religion. By now he had added to the notion of mediated desire the idea that conflict will necessarily arise in every community and that the commonest way to siphon off the internal tension seems always to be the elimination of some surrogate victim, the projection of all hostility on to one member of the group, who is turned into the outsider who carries all the blame and whose expulsion always apparently restores peace and harmony. Paradoxically, since the elimination of the victim achieves a restoration of order within the group, he or she turns out to be a saviour. Thus Girard recovers Rudolf Otto's idea of the holy: the sacred combines the contrary qualities of the *tremendum* and the *fascinosum*: feared monster and blessed friend.

The event which sustains social life more or less harmoniously, then, is periodical bloodletting, literally or figuratively. As Girard notes, some variation on the theme is common in many different cultural traditions. The banishment of Oedipus, the innocent transgressor, saves the city from the plague: he is, as he said he would never be (but not fully understanding his own words), finally driven out as a polluted person in need of expiation (*Oedipus the King*, 402). The killing of Abel, even more curiously, enables Cain to leave the presence of the Lord (but with an apotropaic mark on him) to build the first city (Genesis 4: 8–17). Rome was founded after Romulus murdered Remus. It is as if the historical reality is acknowledged mythologically that every human order is founded on the shedding of an innocent brother's blood. Whatever the variety of particular forms which the practice has assumed in the course of history and across cultures and mythologies, Girard argues that religion has its roots in the internecine and fratricidal conflict, generated in any community by mimetic desire and (it seems) pacifiable and placatable only by resorting to the victimage mechanism. The stabilization of order achieved by scapegoating is of course always fragile and impermanent. The need for sacrificial victims seems insatiable. The process works most effectively when its beneficiaries are ignorant of its true nature. Girard's discovery, when at last he got round to reading the Bible, is that it embodies, more convincingly than any other literature, an exposure of the workings of the victimage mechanism which has the power to free us from the apparently inevitable cycle: a 'revelation' of our true situation which would 'redeem' us, so to speak.

This is a large claim, and perhaps enough has been said already which is vulnerable to criticisms from various quarters to make it seem unsustainable (for example: need Durkheim's theory of religion be the only alternative to Voltaire's?) but, since it is essentially a claim about how to read the Bible, the fairest move now is to show how Girard reads a familiar biblical text.

VI

Briefly, Girard's story is that the gospels in particular (but consistently with the Bible as a whole) offer a remarkable repudiation of socio-psychological violence by revealing the victimage mechanism for what it is and by simultaneously promoting an ethic of love which allows all collaborative readers to rid themselves of the illusion that scapegoating is ineluctable. 'It is always a matter of bringing together the warring brothers, of putting an end to the mimetic crisis by a universal renunciation of violence' (1987, p. 197). Jesus opens up an environment in which all violence (even for self-defence let alone as revenge and reprisal) is unconditionally abandoned:

> You have heard that it was said, 'An eye for an eye and a tooth for a tooth'. But I say to you, Do not resist one who is evil. But if any one strikes you on the right cheek, turn to him the other also; and if any one would sue you and take your coat, let him have your cloak as well (Matthew 5: 38–40).

The problem, so Girard maintains, is that we imagine either that violence is something secondary, which appropriate measures can easily eliminate, or that it is something so ineradicable in human nature that nothing can be done about it. But the gospels tell a different story: 'Jesus invites all men (*sic*) to devote themselves to the project of getting rid of violence, a project conceived with reference to the true nature of violence, taking into account the illusions it fosters, the methods by which it gains ground, and all the laws that we have verified over the course of these discussions' (1987, p. 197). Since internecine violence is generated by the mechanism of mimetic desire, it cannot be interrupted unless it is first revealed as a structural necessity of society (not simply the result of malice), and then firmly responded to by a complete renunciation of retaliatory violence. 'Once the basic mechanism is revealed, the scapegoat mechanism, that expulsion of violence by violence, is rendered useless by the revelation. It is no longer of interest. The interest of the gospels lies in the future offered mankind by this revelation' (1986, p. 189).

Revelation of, and liberation from, the victimage mechanism occur throughout the gospels, according to Girard, in the story of Jesus as a whole but also in the stories he tells and the stories of incidents in which he is involved.

Consider, for example, the weird tale of what happened when Jesus crossed the stormy lake to the country of the Gadarenes (Mark 5: 1–20; Matthew 8: 28–34; Luke 8: 26–39). The story, which appears in three gospels, with interesting though minor variants, opens with the appearance of a completely naked (Luke 8: 27), self-lacerated demon-possessed man who has broken free of the chains with which his fellow townsmen had sought to restrain him and now lives wild among the tombs outside the city. Jesus exorcises him and, since the 'legion' of demons in him begged him not to consign them to the abyss (a place of confinement for demonic forces which, though hostile to God, are ultimately under his control) but to let them enter a large herd of swine feeding conveniently nearby on the hillside. As everyone knows, Jesus allows them to enter the pigs, which then famously rush down the steep bank into the lake and drown. But the strangest feature of this very strange tale is the reaction of the townsfolk who found the man 'clothed and in his right mind', which frightened them, and, when they heard his story, 'all the people of the surrounding country asked [Jesus] to depart from them; for they were seized with a great fear'.

Needless to say, much scholarly and pious ingenuity has been devoted to interpreting this strange tale. It is impossible to reproduce Girard's chapter-length analysis (1986, pp. 65–183). The main point is, however, that the only way to deal with the anomalies in the story is to assume that the townsfolk did not want the escaped lunatic to be rehabilitated and returned to the community. Indeed they need the madman out among the tombs, whom they bound with chains that were not too effective and whom they are now horrified to see clothed and in his right mind. They beg Jesus to leave immediately, presumably because they cannot abide his interfering in their affairs—paradoxically enough, given that he has just succeeded, without any physical violence, in doing what the violence of their chains was supposed to achieve. In reality, then, they did not want the return of the outsider— Jesus' presence among them reveals their true situation. Psychosocially, so to speak, the community apparently needed a demonized member, cast out of the town to squat among the tombs, stripped of his clothing and out of his right mind, punishing his body by gashing it with stones. In effect, to cut short Girard's fascinating analysis, when the man is ready to return among them, the townsfolk realize that they can no longer work out their conflicts and anxieties by projecting them on the tortured existence of the one whom we might as well identify as their scapegoat. This is why Jesus' rehabilitation of the man is greeted not with joy and gratitude, as one might have expected, but with fear and repudiation. The rules that govern the game between the victim and the persecutory community change for good, when the outcast becomes as 'normal' as everybody else. The story shows how hard it is for a community to give up its proclivity for victim-making. 'These unfortunate people fear that their precarious balance depends on the demoniac, on the activities they share periodically and on the kind of local celebrity their possessed citizen had become' (1986, p. 181).

Of course Girard goes into much greater detail than we can follow here; but he has surely hit on the clue which makes sense of a remarkable story. At least he offers a reading of a story which shows his skill as a literary critic. With many other examples he builds up his case that the Judaeo-Christian literature offers a remarkable insight into the workings of the socio-psychological mechanism of sacrificing a victim to absolve everyone else from involvement in, or culpability for, some conflict-riven situation.

VII

The story of Jesus himself, as presented in the gospels, is the story of a man who, although innocent (Pilate: 'I find no crime in him'), was sacrificed to protect the security of the community (John 11: 50), but appeared from the dead, among those who colluded with his betrayal and execution, to proclaim peace (John 20: 19). Jesus, according to Girard's reading of the story, is, of all the victims who have ever been sacrificed, the one who reveals the absurdity of the victimizing mechanism. For once, the victim returned to the scapegoating community, offering them peace and thus the possibility of life without further need for victims. There were witnesses, participants in their way in the violence done to him, who afterwards recovered and were able to recount the event. The violence was incited against him but at last the truth was revealed. All mankind, according to Girard, is caught within the vicious circle, but, on this occasion, knowledge of the mechanism finally breaks through. One man was clear-sighted enough to allow himself to be sacrificed, in extremely brutal circumstances, and, in so doing, disclosed that peace would only come by way of forgiveness. This man becomes a paradigm of non-violence.

For Girard, this means that Jesus is appropriately spoken of as God. 'The authentic knowledge about violence and all its works to be found in the gospels cannot be the result of human action alone' (1987, p. 219). Faced with the only man capable of rising above the violence that had gripped mankind from the beginning we are compelled to acknowledge his divine origins. Girard even endorses the traditional doctrine that Jesus's mother was a virgin (1987, pp. 220–3). The source of his power to interrupt the cycle of the victimage mechanism could only be extra-human. Jesus' response of non-vengeance (cf. Luke 23: 34) breaks the cycle and the sin of his persecutors does not rebound upon its perpetrators. But this is so unprecedented and so unforeseeable an outcome that we might as well speak of a divine intervention. In fact, as Girard ironically concedes, the religion which most educated people regard as fantasy about the supernatural is the very one which reveals something 'natural' about the structural violence in human life and offers a way of understanding and transcending it. In a world dominated by nuclear weapons and industrial pollution it is the Bible which uniquely has the power to free us from the folly of endlessly seeking scapegoats to solve all our conflicts.

Of course Girard is well aware that the Bible is the sacred text of a community, or rather of a group of divided and mutually hostile communities, whose history is marked by heresy-hunting, excommunications and other forms of more or less blatant victimization. But that Christians fail to understand the implications of the story of Jesus only confirms Girard's intuitions about the place of scapegoat-making in any society. For that matter, as he notes on several occasions, the Bible itself is a deeply ambivalent text. Given the amount of killing in the name of God to be found in the historical books of the Bible, and the many executions of dissenters down to the nineteenth century, it seems hard not to admit that a heritage of intolerance and fanaticism is indissociably bound up with the sources and tradition of Christianity. On Girard's view, far from continuing to propagate 'liberal' versions of Christianity which bowdlerize the hate-laden texts and demythologize the strong dogmatic claims, we should get back to reading the Bible, with all its inbuilt violence, in the light of traditional beliefs about the divinity of Christ.

Girard even believes in something like biblical inspiration. 'The gospels cannot be the product of a work that was purely within the effervescent milieu of the early Christians'. On the contrary: 'At the text's origin there must have been someone outside the group, a higher intelligence that controlled the disciples and inspired their writings' (1986, p. 163). While it is not uncommon for modern theologians to adapt traditional themes to their own purposes and sometimes to strain them of their original meaning, it is plain enough here, and often elsewhere, that Girard's beliefs are honest and straightforward in an old-fashioned way. He no doubt recovered his sense of the uniqueness of Christianity from reading novels and studying mythology and social anthropology, but his final position is far removed from any sort of reductionist demythologization.

VIII

The story of the death and resurrection of Jesus is of course at the centre of Christianity. What Girard, as a literary critic in the first place, offers to his readers is a reading of the story—but a reading of the story which, given his interest in sociology and theory of religion, invites us to see how the 'original sin' of any and every society is the making of a scapegoat. Once the nature of this victimage mechanism has been revealed we have at least the possibility of being 'redeemed', of acting otherwise. But Girard clearly believes that the nature of our collective scapegoat-making needed to be revealed—that human beings could never have attained the liberating possibility from their own resources of intelligence and imagination. Girard has come to his understanding of the New Testament from his other studies—he seems in fact to have been surprised to find so much in the Bible when he finally got round to reading it. He speaks now as if Christianity is the sole bearer of this discovery of non-violence as the way to end the injustices of scapegoat-making. Perhaps he might allow that there have been, and are, other traditions of non-violence: Socrates, the Buddha, Gandhi. He certainly allows for the Christian tradition's frequent betrayals of what he takes to be its liberating revelation, as we have already noted; but perhaps he does not allow sufficiently for many other positive elements in the Bible which have little or nothing to do with the victimage system. As Mark Wallace (1989) notes, there is much to question and criticize.

Whatever else may be said, however, it is clear that Girard has opened up an interesting form of 'natural theology'. He starts from the facts of our inveterate disposition towards creating systems of victimage and scapegoating as we seek to preserve social harmony or our cultural identity or whatever. He argues that the Bible offers a privileged insight into the system and enables us, at least in principle, to transcend it. To this extent he is saying that the Bible makes a claim to truth, in the sense that, as we learn to read it, we begin to see the hidden need for a sacrificial victim at work in so much of our social and political life. If the characteristic move in traditional natural theology is from some feature of the world (movement, design or whatever) to the existence of a deity (mover, designer), then we might perhaps say that, in Girard's project, the move is from the facts of victim-making in every known social system to what is revealed about them in the light of the figure of the victim in the Judaeo-Christian narratives. In this sense Girard's approach to the Bible is a form of liberation theology.

But one might accept most of that while balking at his claims that the victimage mechanism, so inveterate and so concealed, required to be revealed by divine intervention. One might, that is to say, find the victim-saviour figure in the New Testament a paradigm—a parable—of a non-retaliatory and non-violent ethic of love and forgiveness while hesitating to believe in his divine origin. But, as Wallace notes, Girard, together with the growing number of literary critics who are currently reinvigorating biblical interpretation, has proposed a reading which seems at least as faithful to the text and as fruitful for social and political action as some more celebrated grand strategies, such as Barth's and Bultmann's, not to mention liberal and liberation theologians.

Note

According to the Oxford English Dictionary, Tindale's version of the Bible (1530) introduced 'the goote on which the lotte fell to scape' (Leviticus 16: 10). The Hebrew transliterated as 'azazel', apparently only half understood, was translated in the Septuagint as 'apopompaios': something which is employed to avert evil. In the Vulgate this became 'emissarius', hence modern French 'émissaire'. In modern scholarship the scapegoat has reverted to the Azazel goat—Azazel being now taken to be the name of some demon (a fallen angel in I Enoch 6: 7)—so that the people's iniquities are carried off by the goat to some demon in the desert. See N. Kiuchi, *The Purification Offering in the Priestly Literature: Its Meaning and Function* (Sheffield Academic Press, 1987).

References

Ashby, Godfrey. 1988. *Sacrifice: Its Nature and Purpose* (London: S.C.M. Press.)

Buckley, Michael J. 1987. *At the Origins of Modern Atheism* (New Haven: Yale University Press).

Byrne, Peter 1989. *Natural Religion and the Nature of Religion: The Legacy of Deism* (London: Routledge).

Galvin, John P. 1982. 'Jesus as Scapegoat? Violence and the Sacred in the Theology of Raymund Schwager', *The Thomist* 46, pp. 173–94.

Gerson, L. P. 1990. *God and Greek Philosophy: Studies in the Early History of Natural Theology* (London: Routledge)

Girard, R. 1965. *Deceit, Desire, and the Novel: Self and Other in Literary Structure* (London: Johns Hopkins).

Girard, R. 1986. *The Scapegoat* (Baltimore: Johns Hopkins).

Girard, R. 1991. *A Theater of Envy: William Shakespeare* (New York: Oxford University Press).

Girard, R. 1987. *Things Hidden since the Foundation of the World* (London: The Athlone Press).

Girard, R. 1977. *Violence and the Sacred* (Baltimore and London: Johns Hopkins).

North, Robert 1985. 'Violence and the Bible: The Girard Connection', *The Catholic Biblical Quarterly* 47, pp. 1–27.

Preus, J. Samuel 1987. *Explaining Religion: Criticism and Theory from Bodin to Freud* (New Haven and London: Yale University Press).

Sykes, S. W. (ed.) 1991. *Sacrifice and Redemption: Durham Essays in Theology* (Cambridge and New York: Cambridge University Press).

Wallace, Mark I. 1989. 'Postmodern Biblicism: The Challenge of René Girard for Contemporary Theology', *Modern Theology* 5, pp. 309–25.

Young, Frances 1975. *Sacrifice and the Death of Christ* (London: S.P.C.K.).

Philosophy vs. Mysticism: an Islamic Controversy

OLIVER LEAMAN*

Islamic philosophy makes a sharp distinction between different categories of believers. Some, and indeed most, believers follow Islam in an unquestioning and natural manner. They adhere to the legal requirements of the religion, carry out the basic rules concerning worship, pilgrimage, charity and so on, and generally behave as orthodox and devout Muslims. Some are more devout than others, and some occasionally behave in ways reprehensible to the teachings of Islam, but on the whole for the ordinary believer Islam presents no serious theoretical problems. There may well be practical problems in reconciling what they wish to do with what Islam instructs them to do, but this for most people is not something which leads them to question their faith as such. It merely leads them to wonder how to reconcile in a practical way the rival demands of religion and their personal wishes.

This category of ordinary believer is very different from the kind of believer who has difficulty in formulating his or her faith in terms which are satisfying to them intellectually as well as practically. Such 'seekers after truth'[1] can follow a number of specifically Islamic routes to try to resolve their difficulties. They may work within the Islamic sciences such as jurisprudence (*fiqh*), law (*sharī'a*) or the rather more controversial discipline of theology (*kalām*). What all these activities share is an acceptance of the basic principles of Islam, which are employed as the basic principles of each individual discipline. If a problem in religious law arises for a Muslim, for example, it can be resolved by applying reason to the principles of Islamic law and relating those principles to the individual issue or case. Although the kinds of reasoning which may be employed by legal theorists and theologians are perfectly acceptable logically, they fall short of the rigour of purely philosophical thought, according to the philosophers (*falāsifa*). Philosophical thought does not work from religious principles; rather, it starts with principles which are themselves demonstratively true, the

---

*I should like to thank Peter Edwards, Irene Lancaster and Michael McGhee for their comments upon an earlier draft. Nothing here should be taken to represent their views.
[1] A frequent expression in Al Ghazālī (1967).

conclusions of yet more abstract reasoning, and so appropriate premises for a reasoning which has as its aim demonstratively and universally
true conclusions. The application of reasoning within religion can only
convince those who adhere to the religion in the first place. The use of
reason within philosophy is designed to convince anyone at all, regard-
less of their religious affiliation.

Now we appear to have three kinds of believer. There are those who
follow their religion without feeling threatened by any theoretical
problems. There are those who follow their religion and cope with a
particular range of theoretical difficulties by resolving them within the
religion through the use of some system of devices designed for that
very purpose. There is yet another group, the philosophers, who follow
their religion and yet try to resolve theoretical difficulties by the application of a kind of Aristotelian logic. A complex machinery of Aristotelian and Neoplatonic logic was developed in the tradition initiated by
Al-Kindī and Al-Fārābī with the aim of dealing satisfactorily with a
whole range of theoretical problems with which theology could not
adequately cope. This approach achieved its most sophisticated treatment in the works of Ibn Rushd (Averroes) who argued that there is
nothing wrong with these differences of approach as between different
kinds of believer. The simple believer, the theologian and the philosopher all believe the same thing, but they believe in different ways. The
philosophers tended to argue that their approach to theoretical problems is really the best, since it is the most abstract and certain, and
subsumes the other theoretical instruments.

Along with this view they adopted a political theory according to
which different kinds of believer should be restricted to their own
techniques of resolving problems. The simple believer, then, is entirely
justified in thinking that God knows everything. The theologian is
entirely justified in trying to reconcile God's knowledge of everything
with the morality of punishing people for committing sins which God
foresaw with perfect certainty. The philosopher is entirely justified in
trying to make sense of the notion of divine omniscience by constructing a very different notion of what such knowledge means as compared
with either the ordinary believer or the theologian. Problems only arise
if the answers provided by a particular category of believer stray into
the wrong territory, so that the theologian thinks that his account of
divine knowledge challenges that of the philosophers, or if the simple
believer considers that his belief is contradicted by the theory of the
philosopher. Once these boundaries of belief start to get confused
disputes inevitably arise and everybody starts to accuse everyone else of
heresy (*kufr*), or at the very least of innovation (*bid'a*). The philosopher would claim that since God has no sensory equipment, it is an
error to suppose that he actually sees what goes on in the world he has

created. Ordinary believers may well have a strong belief, derived in part from their reading of the Qur'ān, that God sees everything which goes on in his world. The philosophers do have an account of how God is aware of what goes on in the world, but it is not the ordinary account. It works from the principle that since he has created the world he is aware of the principles which lie behind that creation, and so can know the abstract principles which structure the very concrete reality of the world. He knows what takes place, but in a different manner from that applicable to his creatures. The philosopher can understand how his account of God's knowledge covers the view of that knowledge which ordinary believers accept, but the ordinary believers themselves probably could not. Were they to be aware of the philosophical theory of God's knowledge, they might come to feel that their faith has been weakened or threatened in some significant way.

What is required, then, is a system of restrictive practices which would restrict particular sorts of arguments to people who fall within different categories of belief. The simple believer would not be troubled with philosophical analyses of his uncomplicated faith, and the theologian would not feel that his approach to theoretical issues is challenged by the wider scope of philosophical thought. This has sometimes been seen as a rather disingenuous move by the philosophers to try to avoid the charge of heterodoxy. After all, the philosophers tended to support theses which seem to contradict religious principles, and they attempt to sidestep this objection by arguing for the distinctness of different forms of reasoning. For example, the philosophers tended to argue for the eternity of the world, against the possibility of individual spiritual or corporeal immortality and against divine knowledge of particulars. They argued that their views here do not really challenge ordinary or theological views, since it is possible to reconcile all such views. But only the philosophers themselves can really understand how this can be achieved, since only they are masters of abstract thought, and it is dangerous to let their weapons fall into the wrong hands. This can result in either a weakening of faith, or an attack upon the orthodoxy of the philosophers. Ordinary believers would get confused were they to be presented with philosophical theories which they could only imperfectly grasp. Even theologians might feel that the philosophers were challenging their understanding of theological concepts, since the philosophers tend to use similar terms to those employed in theology, and the theologians might wish to criticize philosophical theories on theological grounds. This is entirely inappropriate, though, since the different methodologies only have scope within their own fields of demarcation. It is dangerous politically as well as theoretically for practitioners of a particular theoretical discipline to seek to communicate the principles of their discipline to those who are not prepared for it. The notion of an intended *audience* is a very important one in Islamic philosophy and theology, and many important texts are prefaced by reference to a precise (kind of) individual for whom the text is written. This provides an indication of the sort of audience at which the text is aimed, and warns others that this is not the sort of book which is suitable for them.

This might strike us today as the very worst kind of élitism. Why should the ordinary believer not be given the opportunity to experience the possibly more profound understanding of religion which is available to the theologian or the philosopher? Is it really true that the ordinary believer and the philosopher both share the same thing, a common faith, albeit in different ways? Does it not rather seem that the philosopher is taken to have a privileged access to reality, while the ordinary believer has to rest content with a mere simulacrum of that reality? The argument which the philosophers provided goes against this view. The universe is taken to have a rational structure which can be grasped by the use of the logical techniques of philosophy. Once we understand this structure we know why the world is the way it is, how we ought to behave, what lies in store for us after our death, and so on. Strictly speaking, philosophers do not require any divine intervention to achieve such understanding, and nor do they require a religion to help them with such issues. But it is very different for the majority of the population. They do require religion if they are to have an inkling of how the world is and how they are to behave. Religion is a political institution designed to show the majority of the population how to live, and it must use language which is capable of inspiring and persuading ordinary people. The excellence of Muhammad as a prophet is not based upon the fact that God spoke to him, since God spoke to other prophets before him too, but upon Muhammad's political skill in formulating the divine message in such a way as to make it appropriate to the broadest possible audience. Most of his audience would not have understood the message had he presented it in terms appropriate to philosophers. Even theological language would not have impressed most ordinary listeners. What ordinary people need, and what he supplied, is vivid and poetic language, language which can affect the emotions and persuade an entirely general audience.

This political interpretation of religion does not seem to leave much room for the personal relationship between the believer and God. One of the emotional features which might be sought in a religion is precisely the idea that when we pray there is someone there to whom the prayer is directed, someone we can love and who can love us. The idea that this emotional demand is really only appropriate for unsophisticated believers was repellent to many highly sophisticated believers, and led them to reject what they regarded as too abstract accounts of

Islam. Perhaps the most distinguished representative of this reaction is al Ghazālī who arrived in Baghdad in 1091 to take up the equivalent of a chair in Islamic thought in the capital of the sunni Islamic world. His reputation then as now was of an outstanding defender of the principles of Muslim orthodoxy against its traducers. He had previously produced some of the most significant texts of Muslim theology, texts which undertook to refute what he took to be philosophical assaults on Islam and heterodox theological and political views. After a few years he abandoned his position at the summit of the academic hierarchy, gave away his money and possessions and left for Damascus on a prolonged spiritual journey to try to find a form of Islam which could satisfy his longings for a more authentic form of religious experience. He eventually found what he was seeking in Sufism, and in his subsequent writings did a great deal to promote the acceptability of mysticism within orthodox Islam.

This leads us to acknowledge the existence of yet another category of believer, the mystic. Most of the philosophers emphasized the intellectual path to awareness of God. The highest form of happiness attainable is to bring our thinking as closely as possible to divine thought, which involves progressively refining our thought until it is as similar as possible to the rational and abstract principles which underlie the structure of reality. The closer we approach divine thought the less we are ourselves, in the sense that the more abstract and impersonal our thinking becomes. This leads to a version of immortality in which there are no immortal thinkers, just immortal thoughts, since our thinking processes have become identical with that of the active intellect, the repository of abstract thought which structures the reality of our world. This led to a good deal of criticism of the philosophers from the perspective of less elevated thinkers. It would seem that the unsophisticated but devout individual has no route to perfection since he is unable to develop intellectually to the extent required if he is to achieve the highest level of potential human existence. He is either not bright enough, or does not have an interest in intellectual work. The philosophers seem to consign such individuals to a secondary and less elevated form of perfection, that of plodding along in a sort of religious second division far removed from the glories available to their more intellectually inspired peers.

The philosopher, then, seeks to come close to God through the use of reason, through the perfection of his intellectual powers. The mystic aims to feel the divine presence in something like the way in which we taste things. This has implications for the way in which people ought to behave. The philosopher will follow, or at least advocate following, an orthodox religious life-style, while the mystic might well be in favour of an ascetic life-style, the pursuit of Sufi practices and particular techniques to increase spiritual awareness. For example, some mystics try to induce a trance by whirling around in circles. The purpose of such a practice is to imitate the circular motion of the heavenly bodies which were taken to be perfect. A particular feature of circular movement is that it does not have an end, it just goes on and on in an endless and unchanging manner. By contrast, logical reasoning is linear, with a beginning and an end, and proceeds in a step-by-step discursive way. The mystics took this to be a strong argument against the use of philosophy, since God's thought does not come to an end, has no beginning and constantly replicates the principles upon which reality is based. To come closer to God, then, involves going beyond discursive reasoning, by possibly falling into a trance through practices like whirling. Mystics might expect that while in a trance they are able to appreciate more fully the real nature of divine reality than is available through any other method.

A significant contrast between the philosophers and the mystics in Islam lies in the sort of life which they each advocate. The philosophers stressed the importance of living in society for the intellectual development of the individual. Although social and moral virtues are inferior to intellectual virtues, the latter cannot be attained without the former. This is not just the rather banal point that it is easier to do intellectual work if one is adequately fed and gets on with one's neighbours. It is rather the view that a complete life involves the completion of a whole range of activities, not all of the same importance, but all of some importance if one is to live as a human being. Although social life can be a distraction from one's intellectual interests, it is necessarily involved in any accurate grasp of the role of morality and politics in life, and these are themselves appropriate objects of rational enquiry. If we are to live well, we must not neglect the ways in which our lives come into contact with the lives of others. There are dire circumstances in which the philosopher has to live outside society (these are spelt out by Ibn Bājja)[2] but there are great dangers in the individual seeking to separate himself from his fellows. He may come to think that he is more important than he really is, or that he is more distinct from the rest of humanity than is really the case. He may think that he has come closer to God than other human beings. Living in society with other people helps prevent us from falling into self-aggrandizement.

The philosophers would argue that the mystic is in danger of falling into a variety of errors. The latter may believe that human beings can live well in exclusion from others. He may believe that the customs of religion are unimportant. He may even confuse contact (*ittiṣāl*) with God with union (*ittiḥād*) with God. The mystic is using language

[2] This is clearly argued in Rosenthal (1958).

pushed to its limits when he describes his experience. He quite self-consciously does not want to use the rules of a rationally limited language. The experience which he achieves has been so vivid and direct, while its description by contrast seems so tawdry and secondhand. The mystic is like the philosopher in that he too rejects *taqlīd*, blind obedience to tradition, but very different in so far as he tries to construct a unique route to God, one based upon the formation of an emotional relationship which brings him nearer to experiencing the deity. It might appear that the mystic comes closest to the ordinary believer, in that they both expect some emotional result from their adherence to Islam. If we think about the account which the philosophers gave of the purpose of religion, to communicate widely the truths of philosophy, then the emotional aspect of faith will hardly be a minor matter. The prophet uses his political skill to elicit an emotional response from his audience, and he uses a whole range of rhetorical, poetical and other persuasive techniques with very little in the way of logical value. Ordinary people are unable to regulate their lives satisfactorily through the use of reason alone, they require in addition an emotional reason for following religion and are entitled to expect certain kinds of emotional consequences in return.

Take the example of the after-life, for instance. For the ordinary Muslim, the after-life might be seen as a place where he will be rewarded or punished in a physical sense in proportion to his virtues and sins in this life. For the philosopher the notion of the after-life is a symbol of the persisting consequences of our actions[3] and is certainly not to be regarded as a heaven or hell with houris and torture as features. Even theologians reject the ordinary notion of an after-life as a locus for physical events.[4] Yet it might be thought that the ordinary believer needs the emotional support which is to be found in the idea of a personal physical reward and punishment system if he is to behave well and avoid sin. The mystic then resembles the ordinary believer in seeking emotional satisfaction as a religious end, while also resembling the philosopher and the theologian in following a rational theory of some complexity which explains how that end is to be attained. It seems on these grounds that it will be difficult to debar the mystic from presenting a perfectly acceptable view of a route to God. Since one of the excellences of Islam is taken to be its ability to allow a wide range of routes to God, the mystic is a difficult believer to exclude.

Yet the philosophers sought to exclude mystics from the community of Islam. Surely an important reason is the apparent desire of the

[3] This concept is explored in more detail in my books on medieval philosophy in the Islamic world: see Leaman, 1985, 1989, 1990.
[4] See in particular Al Ghazālī (1989).

mystics for a completely different account of the relationship between religion and experience. The mystic places great importance upon *private* experience, upon the privacy of the individual's religious experience. Yet can a private experience serve as the foundation of religion? The account of religion which the philosophers provide emphasizes its publicity, its accessibility to the widest possible category of human beings. Now, it might be said that mystical systems are also accessible to the public, and have been popular and enjoyed many followers since they were established. Yet the difficulty is in knowing what it is that people follow such mystical movements *for*. If what they are seeking is a less materialistic life-style, then they can be catered for within the confines of traditional Islam. If, however, they are looking for an experience which is going to underpin their whole religious faith, they are looking in the wrong direction. We know from the accounts which mystics provide of their experiences that they are difficult if not impossible to communicate to others, and there are dangers in basing a whole religion, a way of life, on an experience which may or may not occur, and which is difficult to describe.

Yet if we think about Al Ghazālī's concerns nine hundred years ago we are faced with the complaint that he felt entitled to an emotional relationship within Islam, a relationship which he could not find through theology or philosophy. Something was missing from his religious life, and he sought to find it, as many have in the past and continue to do today, in a more mystical interpretation of his faith. The philosophers would argue that there is nothing wrong with pursuing an emotional relationship provided that one does not place more weight on it than it can bear. We can see why this is wrong if we take a different kind of emotional relationship, albeit one which the mystics use a good deal, the relationship involving love. When one is in love one may well have a whole series of deeply felt emotional experiences towards the object of one's love. One may enjoy times of bliss and even physical requital of one's passion. On the other hand, one may not. The form which love takes for one individual may be very different from its form for another. Some people may experience love through the performance of perfectly ordinary tasks which have as their object the benefiting of the loved person without any strong feelings going along with the actions. Indeed, throughout a protracted relationship there may in the end be no strong feelings left at all, and yet there might be no difficulty in describing the relationship as one based upon love. If love can be described as an institution, and it does possess some of the structure of an institutional relationship, then part of what gives it its force are the actions which people perform from which it can justifiably be claimed that love is present. If I claim to love someone, but refuse to give any evidence of my feelings even when the context calls for and permits such evidence, it might well be doubted whether I do really love someone. As with faith, love can take many forms, and not all of these forms depend upon the presence of some strong emotional relationship being experienced at all times of the relationship. We might say of someone who insisted on identifying love with a particular set of emotions that this was a very narrow view, a view which might be unlikely to survive the early stages during which one's ardour is at its greatest strength.

The philosophers suspected that the demand for emotional satisfaction made by the mystics was an excuse for self-indulgence. The mystic wants more from religion than it can realistically be expected to provide. Al Ghazālī (1967) reports:

> I apprehended clearly that the mystics are men who had real experiences, not men of words, and that I had already progressed as far as possible by way of intellectual apprehension. What remained for me was not to be attained by oral instruction and study but only by immediate experience and by walking in the mystic way.

Yet there already exists a type of emotional satisfaction in the experience of performing one's religious duties conscientiously, and perhaps that is what one should expect the emotional aspect of religion to be. Going further and seeking through special practices to extend one's emotional religious range risks falling into all kinds of dangers. One may end up identifying oneself with God, separating oneself from society and foregoing the ordinary religious obligations which are so important a part of the life we ought to live. Along with these personal dangers there exists a conceptual danger, that of identifying the private mystical experience with the meaning of religion. Such an identification would suggest to those not able or uninterested in attaining such experience that they were not really part of the religious community at all. Mysticism can result in limiting the scope of religion to only a few people, whereas it is the whole purpose of religion to widen the possibility of sharing in knowledge and practice to as many people as are available to hear the message. What lies at the heart of religion, then, cannot be some sort of experience, however important and impressive in nature, since that would be to exclude the majority of the population from religion.

This might appear to be a poor argument. Could not everyone endeavour to attain that experience and thus come to share in the most important part of religion, the emotional part? This is certainly a question which the mystic would like to pose. Yet asking everyone to become a mystic is like asking everyone to become a philosopher or a theologian. Not everyone can, nor would everyone want to have that approach to their faith. The mystics themselves describe at some length the difficulties of following their path. But if the meaning of religion is something which most people cannot attain, then religion cannot be for most people, which given the account of religion provided by the philosophers is an absurdity. Religion is important precisely because it can attract and encourage the mass of people to take correct decisions about their lives, and its meaning must be a public and accessible notion. The mystic thus reveals his misunderstanding of the fundamental nature of religion with his exaggerated emphasis upon the importance of religious experience.

Any view of religion which is defined in terms of its public character is going to find it difficult to accommodate the mystic. However determined Islamic philosophy was to accept a wide variety of routes to God, the mystical route was always treated with some suspicion. Behind the rational arguments of the philosophers against mysticism lies a powerful emotion. For the philosophers the nature of reality is essentially knowable, and there is no reason in principle for everything about reality not to be apprehended. Since we are finite beings we are unable in practice to grasp it all, but we can understand the very general rational principles which form the framework upon which all the details of reality hang. Mystics do not necessarily share this assumption. They may well feel that behind what can be known lies a realm of reality which is essentially ineffable, and only partially available to us if at all. Mystics are attracted by the idea that there is a potential mystery in why things are as they are in the world. After all, they follow the mystical route once they have mastered the philosophical route, and remain feeling unfulfilled. Philosophers not only try to demonstrate the rationality of the organization of the world but also want to show that we should be satisfied once we have that information. The protracted dispute between the mystics and the philosophers in the Islamic world reveals a good deal about the differing natures of the enterprises involved. The dispute centres upon the role of experience in religion and has far wider relevance to the philosophy of religion than might at first appear to be the case.

References

Al Ghazālī. 1967. *Munqidh min al-dalāl* ('The Deliverer from Error'), trans. W. M. Watt. (London: Allen and Unwin).

Al Ghazālī. 1989. *The Remembrance of Death and the Afterlife (Kitāb dhikr al-mawt wa-mā ba'dahu)*. Book XL of *The Revival of the Religious Sciences (Iḥyā' 'ulūm al-dīn)*, trans. T. J. Winter. (Cambridge: Islamic Texts Society).

Leaman, O. 1985. *An Introduction to Medieval Islamic Philosophy* (Cambridge: Cambridge University Press).

186

Leaman, O. 1988. *Averroes and his Philosophy* (Oxford: Oxford University Press).

Leaman, O. 1990. *Moses Maimonides* (London: Routledge).

Rosenthal, E. 1958. *Political Thought in Medieval Islam* (Cambridge: Cambridge University Press).

Non-Conceptuality, Critical Reasoning and Religious Experience. Some Tibetan Buddhist Discussions

PAUL WILLIAMS

The Dalai Lama is fond of quoting a verse attributed to the Buddha to the effect that as the wise examine carefully gold by burning, cutting and polishing it, so the Buddha's followers should embrace his words after examining them critically and not just out of respect for the Master. A role for critical thought has been accepted by all Buddhists, although during two and a half millennia of sophisticated doctrinal development the exact nature, role and range of critical thought has been extensively debated.[1] In general doctrinal difference in Buddhism has been seen as perfectly acceptable, reflecting different levels of understanding and therefore different stages on the path to enlightenment. Buddhism has tended not to look to or expect doctrinal orthodoxy, although there has always been a much stronger impetus towards ortho*praxy*, and common (largely monastic) code and behaviour has perhaps played a comparable role in Buddhism to common belief and creed in some other religions.[2] Nevertheless an acceptability of doctrinal divergence has not lessened the energy and vigour devoted to lengthy and sometimes fiercely polemical debate between teachers and schools. This was nowhere more so than in Tibet, where doctrinal debates—sometimes lasting all night—to the present day form the central part of a monastic education in most of the largest Tibetan monastic universities.[3]

From the beginning the Buddhist tradition has characterized enlightenment as 'seeing things the way they really are' (*yathābhūtadarśana*), a seeing which differs in some crucial way from a perception of the way things appear to be to the unenlightened

[1] I have further discussed this theme in a different context in Williams, 1991.

[2] For the importance of these points in appreciating certain major developments in the history of Buddhist thought from about the second century B.C.E. onwards, associated with the rise of Mahāyāna Buddhism see Williams, 1989, ch. 1.

[3] Since the Chinese takeover in 1959, and subsequent destruction of nearly all Tibetan monasteries, these great monastic universities have been re-established in India, particularly in Karnataka.

person. This gap between appearance and reality of course raises essentially philosophical questions, even if they are embedded in a wider framework which we are pleased to call religious and which involves characteristic Buddhist forms of psycho-physical practice. Rivals in debate, even one's co-religionists, do not see things the way they *really* are. In the final analysis their perspective falls short of the complete path to enlightenment, and compassion requires discussion. What we would call philosophical investigation and understanding has always been characteristic of Buddhism, and we know that Hindus too saw one of the characteristics of the Buddha himself as his employment of reasoning and logic in order to question the traditions of Brahmanical orthodoxy.[4]

Yet 'seeing things the way they really are' is an unclear and ambiguous way of speaking. Suppose I am a latter-day Sherlock Holmes, and I reason through genius and perhaps a little gentle experiment that Archibald is the thief. We have here a case of *knowing that* Archibald is the thief; I see that really it is Archibald who is the thief and not Jemima as everyone including myself previously thought. This mode of seeing things the way in which, in this context, they really are clearly differs from the case of my happening to see Archibald stealing the suet pudding with my very own eyes. Yet either mode could be said to be 'seeing things the way they really are'; either could contribute to a conviction in a court of law, and neither need imply the other. Buddhist thought has always evinced a tension, sometimes manifest in institutional differences, between the claim to know that something is the case through critical analytic investigation—as mental events go, a perfectly normal sort of event—and the knowing of the way things really are which seems to accompany or is said to be identical with a particular sort of direct experience, a gnosis claimed to be incontrovertible and identified as an essential experiential element in what is called 'enlightenment', in other words, a *para*normal mental state. The problem which rears its head again and again in Buddhist thought, and provides a thread running through all the immense literature of Tibetan Buddhism, is that of determining the exact relationship between these two 'modes of knowing' given that the knowing which proceeds from critical investigation must by virtue of its linguistic basis require the use of concepts, while the highest form of experiential knowing, our paranormal mental state, the *sine qua non* of enlightenment, is held in some important sense to be direct, non-linguistic and—as it is usually expressed in translations into English—non-conceptual.

[4] See, for example, the Hindu myth of the origins of the Buddha's appearance on earth in the *Viṣṇu Purāṇa*, trans. in O'Flaherty, pp. 231–5.

The problem was there from the very beginnings of Buddhism in Tibet. We are told that in the eighth century C.E. there took place a great debate (or series of debates) in the presence of a Tibetan emperor. The protagonists were an Indian teacher of the school of Madhyamaka Buddhist philosophy named Kamalaśīla, and a Chinese monk called Mahāyāna, who seems to have been a follower of some form of Ch'an.[5] The differences between Kamalaśīla and the monk Mahāyāna are often discussed in later Tibetan literature, almost invariably to the disparagement of Mahāyāna who is agreed by Tibetan sources to have lost the debate. And yet it is important here to see what they have in common. Mahāyāna begins by telling us that everything is generated by a *vikalpa* of the mind (*thams cad sems kyi rnam par rtog pas bskyed pas*).[6] The word *vikalpa* is one of a range of Sanskrit words associated in Buddhist thought with construction, usually conceptual and linguistic, and with the falsification of what some Buddhist writers (but not all) would claim is given immediately in preconceptual experience. *Vikalpa* in particular is associated with discrimination in terms of binary categories x/not-x, a reification of experience in terms of opposition, an hypostatization which is thought to create the framework for suffering which for all Buddhists is the very nature of unenlightenment—*saṃsāra*—and arises in part at least from our inveterate mental tendency to grasp and attempt to hold that which naturally changes.[7] When Mahāyāna says that *all* arises from this dichotomizing discrimination he undoubtedly means all *saṃsāra*, all unenlightenment.[8] Enlightenment therefore lies in reversing the binary hypostatizing operations of the mind. Kamalaśīla does not disagree with the primacy of binary hypostatization in the process of unenlightenment. He states quite clearly that 'that absence of dichotomizing discrimination is the Dharmadhātu—Ultimate Expanse—which is the essence of all phenomena' (*chos thams cad kyi rang bzhin chos dbyings rnam par mi rtog pa de*: Houston, 1980, p. 19). In other words (ignoring for the moment an imprecision which would make a mental state here *itself* the ultimate truth) a mental state where there is no such discrimination is indeed in

[5] Better known by its Japanese name of 'Zen', although Mahāyāna's Ch'an should not necessarily be identified with any form of Zen found in Japan, past or present.

[6] From the *mKhas pa'i dga' ston* of dPa' bo gtsug lag. Text supplied in G. W. Houston, 1980, p. 18.

[7] On the meaning of *vikalpa*, and the whole range of other 'construction terms', see Williams, 1980.

[8] As is stated in the *Bhavasaṃkrāntiparikathā* verse 5, attributed to the great Indian scholar Nāgārjuna: *'jig rten rnam par rtog las 'byung*. *'Jig rten* here refers to *loka*, the world, but it is undoubtedly *loka* in opposition to supramundane (*lokottara*), i.e. world as unenlightenment. See Sastri, 1938.

Paul Williams

some fundamental sense identifiable with—or constitutes an essential characteristic of—enlightenment. But Kamalaśīla shrinks from drawing the conclusion which Mahāyāna is reputed to have derived from his initial starting point—that enlightenment apparently lies in the cutting of all mental activity, not to mention moral activity or religious practice: 'Whoever does not think anything, does not do anything—that person will be completely liberated from unenlightenment' (*gang zhig ci la yang mi sems zhing ci yang mi byed pa de 'khor ba las yongs su thar bar 'gyur ro*). Mahāyāna means what he says. One should not think anything (*ci la yang mi sems*), not examine anything (*ci la yang mi rtog*), and not investigate anything (*ci la yang mi dpyod*). Enlightenment is a mental state in which there is no mental act, and thus *de facto* the acts which engender unenlightenment are eliminated.[9]

Mahāyāna's antipathy towards the binary hypostatization of dichotomizing discrimination is just one aspect in Buddhist thought of a frequently expressed unease about the way our minds make sense of the world of incoming sensory data, a making-sense which is also thought to create a basis and framework for the enslavement and suffering which flows from misunderstanding the nature of things. A further and broader category of mental act often viewed with antipathy is that of *saṃjñā*, and our Tibetan sources also see Mahāyāna's conclusions arising from his advocacy of a paranormal mental state attained through meditation where *saṃjñā* has been negated (Houston, 1980, text p. 25). *Saṃjñā* is said to produce dichotomizing discrimination. I have discussed this notion of *saṃjñā* at great length elsewhere.[10] To summarize: The term *saṃjñā* designates the mental act of apprehending *x* to be a case of '*x*'. That is, in the case of the perception of a blue object, an object is seen to be qualified by the sign (*nimitta*) blue, and this is capable of being linguistically articulated by the statement '*x* is blue'. The *saṃjñā* is the mental act which sees *x* to be a member of the class 'blue objects' as a result of the apprehension of a sign, which stands as a sign of class-inclusion. It is a 'seeing that' which is not identical with, but is bound up with, linguistic articulation (the sign here is not a

[9] Note that I am not concerned here with whether the monk Mahāyāna *really* held the 'blank mind' thesis. There is some evidence from relatively recently discovered Central Asian material that his position may have been rather more subtle than this. It may also bear some relationship to the Chinese Taoist notion of 'doing nothing' (*wu-wei*) which, if I understand it correctly, does not always entail literally doing nothing at all. But Kamalaśīla and most later Tibetan writers certainly thought Mahāyāna held the blank mind thesis. This thesis is criticized by Kamalaśīla in his three *Bhāvanākramas*, and there can be no reasonable doubt that it is there the monk Mahāyāna who is being criticized.

[10] See the discussion in Williams, 1980.

linguistic sign, but an actual occurrence of, in this case, blue, known through perception). The apprehension of the sign blue as a sign of class-membership is already felt by many Buddhist thinkers to involve a degree of falsification, since Buddhist thought denied the fundamentally real status of universals which were held to be at variance with an appreciation of impermanence. To see x as a member of a class is to be one stage removed from an appreciation of x's actual uniqueness, a true uniqueness which is usually held in Buddhism to be radically impermanent and the reification of which—partly the result of a projection of class-membership which is mutually implicated with linguistic articulation—leads to expectations (of endurance, satisfaction, etc.) which are bound eventually to be disappointed.

If a *saṃjñā* is a mental act involving class-inclusion the result of which is capable of being represented in linguistic form as, paradigmatically, a subject-predicate sentence, then the act of *saṃjñā* would appear to involve, and perhaps to correspond quite well with, what is usually thought of in philosophical circles as 'conceptualization'. The vagueness of the the word 'concept' in modern philosophical writing has become as notorious as the frequency of its use, a vagueness seen as a virtue by Peter Heath but bemoaned as the cause of a lamentable imprecision and rampant misunderstanding by C. W. K. Mundle.[11] According to Heath in general one is said to have the concept x if one knows the meaning of the word 'x'; one can pick out or recognize a presented x, or think of x's when they are not present, and/or if one knows the nature of x, that is, if one has grasped the properties of x which make x's what they are (Heath, 1967, p. 177). The common denominator between all these conditions for having the concept is an ability to see x as a member of the class 'x's' It is commonly, although not exclusively, thought to involve linguistic competence. Thus, as Geach points out, 'if someone knows how to use the English word 'red', he has the concept of red'.[12] Paradigmatically, although not exclusively, we might say that someone has the concept of red if he or she can correctly attribute membership within the class of red things. In Buddhist terms here, one has the concept expressed by a term if one is capable of attributing the sign (i.e. the occurrence of red) which is referred to by that term to a subject which is characterized by that sign. For Buddhist writers all *saṃjñā* involves conceptualization and, it seems, all conceptualization involves *saṃjñā*. If *saṃjñā* is problematic for Buddhists then so, of course, is conceptualization. And although a Chinese Buddhist, Mahāyāna is following an old Indian Buddhist precedent when he sees enlightenment as being a mental state antitheti-

[11] See P. L. Heath, 1967; and C. W. K. Mundle, 1970, 1:8.
[12] Peter Geach, 1971 reprint, p. 12.

cal to *saṃjñā* and therefore the conceptualizing and linguistic process. In the *Kāśyaparivarta*, attributed to the Buddha himself, it is said that a monk who has destroyed conceptualization (*saṃjñā*) and dichotomizing discrimination (*vikalpa*) is liberated, he has nothing more to do.[13]

Mahāyāna and Kamalaśīla both appear to agree that enlightenment involves a mental state free of conceptualization and therefore for the person undergoing it ineffable, and apparently free of any other form of hypostatizing or discriminating mental act. Where they disagree is in terms of what this actually amounts to, particularly in terms of the process by which it is brought about. For Kamalaśīla a mental state free from conceptualization cannot be the result of simply ceasing to think, of making the mind a blank. There are paradoxes involved in any claim to have a completely empty mind, and the subsequent claim that the mental state was non-conceptual. For Kamalaśīla one could never bring about such a state. He comments that anyone who thinks that they will not think of anything is in fact doing a great deal of thinking (Houston, 1980, p. 19)! Of course, even supposing one could bring about such a state, if a mental state is genuinely empty of content then it is difficult to see how one could call this a mental state. There cannot be an *experience* lacking content, for there could be no way of distinguishing it from no experience at all. If one really had a blank mind, one could have no grounds for claiming that one's mind is or was a blank, or that it is or was in a non-conceptual state (assuming one intends to mean by 'blank mind' and 'non-conceptual state' more than the mental state of a stone, i.e. no mental state at all). One is tempted to assert that such an absence of experience could have no significance, religious or otherwise. But this would, I think be to overstate the case. It could have no significance for someone in that state, that is, *in itself* it is just a blank and therefore insignificant. But for one who is not enlightened it could be said to be significant as that towards which one strives, that is, the cessation of the forces which lead to unenlightenment, and indeed this is perfectly coherent, given the premises of the monk Mahāyāna. If all mental activity is the cause of unenlightenment, then a cessation of mental activity would be enlightenment.[14] Whether that could be worth striving after, a worthy goal of religious life, depends on one's estima-

[13] See A. von Stael-Holstein, 1926, section 136.

[14] Actually, the matter is a bit more complicated than this, since in Buddhist thought mental acts associated with the forces which produce unenlightenment generate seeds which will normally contribute towards a perpetuation of the process of unenlightenment into further births. Thus cutting all mental processes might prevent *further* seeds of unenlightenment, but it would not in itself destroy those seeds which are already present 'in the mental continuum'. This also is behind Kamalaśīla's criticism that simply making the mind a blank would not have the force to generate liberation.

tion of unenlightenment and human potential. But clearly, as Kamalaśīla points out in his reply, it would not be the state of perfect wisdom and compassion, characterized by attributes such as remembering all one's infinite past lives, perfect altruism for the benefit of others and so on, which are said to accompany the attainment of the highest goal in Buddhism.

The fact that the significance of a blank mind could only be for those who had not yet attained it is important, for it is just the process of attainment that Kamalaśīla sees as providing the most powerful argument and coherent basis for his claim that a mere blank mind—a *tabula rasa*—is not what is meant by the non-conceptual state which is associated with enlightenment. His argument is in terms of how the non-conceptual state is brought about, but it is clear that it also involves what it is to be a non-conceptual state in the significant sense referred to in the context of generating enlightenment.

Kamalaśīla begins his attack on Mahāyāna by pointing out that simply making the mind a blank is the very antithesis to the wisdom which results from the correct analysis of things (*so sor rtog pa'i shes rab*=Skt.: *pratyavekṣaṇaprajñā*), which is to say, it is the very opposite of seeing things the way they really are. In other words, for Kamalaśīla critical analysis is not to be denied, but is the constituent, an essential constituent, in the way by which we come to understand things and thus eventually attain a mental state which while it is non-conceptual is clearly not a mere blank. Analytical wisdom is the very root of correct gnosis, Kamalaśīla comments. To abandon that is to abandon any supramundane gnosis. Without the wisdom of correct analysis how could a *yogin* in his meditation attain to the mental state which is not subject to dichotomizing discrimination?[15] A mere blank is not able to be a cause for bringing about a genuine state of non-conceptuality (Demiéville, 1952, p. 350). That is, knowing that something is the case, a 'knowing that' reached through critical analysis, is a prerequisite to subsequently attaining through meditation a mental state of direct acquaintance with the state of affairs which was previously known only through reasoning. Kamalaśīla may accept a mental state which is in some sense non-conceptual, but the process by which it is brought about implicitly involves the use of conceptual reasoning. Moreover although that eventual state may have some similarities to the blank mind referred to by Mahāyāna, in that they both purport to be

---

[15] I am basing my discussion here on the account of Kamalaśīla's position outlined in the *mKhas pa'i dga' ston*, for which I have the Tibetan text reproduced by Houston. However, this account follows very closely that written by Kamalaśīla himself in his third *Bhāvanākrama*, translated by Etienne Lamotte in the Appendix to Demiéville, 1952.

non-conceptual, in actual fact there is a great deal of difference between them. For Kamalaśīla the process of attainment in some sense determines the eventual state. It may be non-conceptual but it is a non-conceptuality which contains, as it were, all that went before. It follows that for Kamalaśīla while the mental state itself may be non-conceptual in that a person enjoying such a state is not engaging in conceptualizing, it is possible for others, or the *yogin* after his gnosis, to adequately explain what the state was, and its content. For this state has some content, and is thus distinguishable from the blank state which in its very absence of content could not be said to be an experience at all. A distinction is to be drawn between a mental state which is non-conceptual and the contents of that state, that is, what the state is of, expressed in terms of who the subject of that state is and what it is directed towards, a direction determined by the analyses which had taken place previously. The monk Mahāyāna assumed that there is a paradox in using conceptual means to bring about a non-conceptual state. For Kamalaśīla this is an unwarranted assumption.

We cannot begin to appreciate the force of Kamalaśīla's counter-argument to the blank mind unless we appreciate what Kamalaśīla's Madhyamaka philosophy, and those Tibetans who came after him, mean when they refer to the 'ultimate'—that the knowledge of which forms the referential content of the ultimate gnosis. According to a widely accepted etymology of 'ultimate' (*paramārtha*) it is called 'ultimate' because it is the supreme (*parama*) referent (*artha*).[16] It is, our author says, the referent of supreme gnosis free from dichotomizing discrimination. Thus the mental state associated with direct non-conceptual gnosis seems to *have* a referent, even if there is no sense in the experience itself that 'This is the supreme referent.' It is perfectly possible for me, in seeing blue for example, to be so absorbed that I am not aware *that* I am seeing blue. Yet it does not thereby become false that I am seeing blue, that blue is the object of my experience.

This view appears to be at variance, however, with a verse much quoted in later Tibetan literature from the eighth century Mādhyamika Śāntideva: 'Reality' (by which he means the ultimate), Śāntideva says, 'is not a referent of the mind.'[17] This point was taken up in Tibet by one of the early translators, rNgog Lotsawa, and was particularly associated with the name of one of the greatest Tibetan scholars, Sa skya Paṇḍita Kun dga' rgyal mtshan (pronounced: Kern ga gyel tsen—1182–1251): 'Since the ultimate is free from *prapañcas*, it is not a referent of

[16] See Bhāvaviveka's *Tarkajvālā*, quoted in Tsong kha pa's *Drang nges legs bshad snying po*, p. 31.
[17] *Bodhicaryāvatāra* 9:2, ed. P. L. Vaidya, 1960: *buddher agocaras tattvam*.

conventional usages like "existence", "nonexistence", "negation" and "proof". This is because it is not a mental referent.'[18] The word *prapañca* refers to yet another dimension of the process of conceptualization, discursivity and hypostatization, this time the whole process of conceptual proliferation involving or through linguistic reification.[19] The ultimate is beyond *prapañcas*, it is ineffable and the experience of it is non-conceptual because, Sa skya Paṇḍita is saying, it is not a mental referent at all. In saying that the experience of the ultimate is non-conceptual we are also saying not only that there is no conceptual activity taking place while a subject is undergoing the experience, but also it is *essentially* non-conceptual in that it lacks all objective content. It is not clear here whether Sa skya Paṇḍita also wants to hold that this experience lacks a subject, although given the normal correlation of subject-object dichotomy in Buddhist thought it is almost certain that he would want to maintain this. If so, then once more if it is taken literally it is difficult to see how this could be called an experience at all. If it is not possible *for anyone* to say that *x* is having the experience, or that it is an experience of *y*, then it is difficult to see what it means to speak of an experience here, and Sa skya Paṇḍita's position in fact collapses into that of the monk Mahāyāna.

The view that a gnosis directed towards the ultimate has no object and therefore lacks content was opposed in Tibet with considerable vigour and hermeneutical skill by some of the great scholars of the dGe lugs school (pronounced: Geluk), founded by rJe Tsong kha pa in the late fourteenth century. In his massive treatise on the ultimate, spoken of as emptiness, Tsong kha pa's pupil mKhas grub rje (pr.: Kay drup jay) writes that such an interpretation is obviously absurd, for it would follow that the Buddha himself taught the ultimate truth without knowing it—since there is nothing to know! Moreover since on this basis there *is* no ultimate truth, because if there were it could presumably be cognized as an object of the mind, so there ceases to be any distinction between the way things appear to be and the way they actually are.[20] In other words, if the ultimate is not a mental referent there cannot be an experience of the ultimate. From which it follows that for mKhas grub rje one can only meaningfully speak of an experience if it has a referential content. Thus it is quite clear that Śāntideva should not be taken to mean literally that the ultimate is not an object of the mind, that is, it cannot form the objective content (the 'intentional referent', in Brentano's sense) of a mental state of supreme gnosis which

[18] From the *mKhas pa rnams 'jug pa'i sgo*, Tibetan text quoted in Jackson, 1987, p. 396, note 95.

[19] See once more Williams, 1980.

[20] mKhas grub rje dGe legs dpal bzang, 1972, p. 430.

takes the ultimate as its referent. mKhas grub rje comments that 'as for the nature of the ultimate truth, the referent which is an actual mode of being that is not the referential sphere of a mind which is deluded by dualistic appearances is the ultimate'.[21] In other words it is possible to talk of a mental state as both non-conceptual and having a referential content. The claim that it has no referent at all should be taken as signifying that its referent is not apprehended in the way of ordinary dualistic experience. What mKhas grub rje does here is direct attention away from the primacy of the non-conceptual experience itself, a primacy which had been taken to suggest paradoxical philosophical, that is conceptual, conclusions. mKhas grub rje in fact draws a distinction between the experience, which may be non-conceptual in the sense that it does not involve any apparent conceptual activity while occurring, and may have been brought about precisely by discovering the range and limitations of conceptuality, and the structure of that experience revealed to others, or the same *yogin* in his post-meditational state. It is at least not obviously paradoxical to claim to have had an experience which has a subject-object structure even though at the time the experience was undergone no subject or object were consciously felt to be present. There is a distinction between an experience and that which is experienced. That the experience here is non-conceptual is vouchsafed by the nature of the experience itself and what led up to it. None of this suggests that it *cannot* be conceptualized. In all of this, I think, mKhas grub rje and the dGe lugs tradition are making good philosophical sense and going some way to avoiding the problems associated with non-conceptuality that bedevilled the monk Mahāyāna and even touched Sa skya Paṇḍita.

This may be an appropriate point to summarize what I think I am saying here in general about a non-conceptual experience of the ultimate (whatever that might be). I can make sense of an experience which when it is occurring does not appear to involve the conscious use of any conceptual categories. What I do not think this implies is a conclusion that the experience is non-conceptual*izable*. If this point requires that there must be some sort of subconscious conceptualization going on, then so be it. If to have a concept of x involves an ability to pick out or recognize a presented x then it seems to me that this can occur at a level subliminal to ordinary focused awareness. There can thus be different levels of conceptualization. To have the concept of x also includes an ability to use x-terms (if they are used at all) correctly. It does not require that x-terms are actually employed on every possible occasion. An experience which is non-conceptual in the sense that it could *never*

---

[21] Ibid. p. 431: *don dam bden pa'i ngo bo ni / gnyis snang 'khrul pa'i blo yi spyod yul ma yin pa'i gnas lugs kyi don ni don dam pa yin la /*

be conceptualized even in subsequent cool reflection is simply mean-ingless. It could not be meaningfully described as an experience, or of the ultimate. To a claim to have had a non-conceptual experience of this type one can reply that it is a contradiction and anyway, so what? *This* experience (supposing one grants that it is an experience) could not be the experience claimed to have such significance in religious discourse. The monk Mahāyāna's blank mind can be made meaningful only because of his apparent contention that *all* mental activity is unenlightenment. The blank mind gains meaning structurally, as not-unenlightenment. But we have seen that the blank mind is nevertheless still incoherent as an explanation of the experience of Buddhist enlightenment.

I can also understand that someone may enjoy an experience for which no description or explanation is felt to be adequate. This very inadequacy requires that the original experience was in some sense conceptualized or is conceptualizable, otherwise one could not say that the description is inadequate. Likewise I can understand a claim that a particular person or group may lack conceptual dexterity, or a symbol system may be too impoverished to adequately symbolize the experi-ence involved,[22] although this would seem to require a second symbol system in which it *could* be at least more adequately symbolized, for otherwise how again would we know that the first symbolization was inadequate?[23] But none of this warrants the claim that the experience is of such a type that it could never be conceptualized, in the sense in which I have spoken of conceptualization. It is the issue of the non-

[22] For an ingenious example, see here Henle, 1970. Henle's point is that we can construct a symbol system in which something which could be expressed perfectly adequately in our normal English symbol system could not be said in the new system without paradox. It thus becomes ineffable in symbol system (i). It would follow, of course, that the claim of ineffability is relative to a symbol system. It ceases to involve *inherent* ineffability, or non-conceptualizability.

[23] Of course, the notion of *adequacy* depends on context and purpose. Adequacy in describing an experience does not mean literally giving the hearer the experience through one's use of words. A lot of so-called 'mystical' writing on ineffability often involves no more than the claim that having the experience is better than, and contains features not contained in, simply hearing about the experience. If the claim that the experience is beyond concepts involves simply differentiating between speaking about the experience and actually having it, then this is non-controversial—but as mKhas grub rje realizes, this need not entail such radical and absurd epistemological conclusions as a claim that the ultimate cannot be an intentional referent, or indeed spoken about in a way which enables the discourse to be inserted into a spiritual-cum-philosophical system.

conceptualizability of the non-conceptual experience which is approached both here and elsewhere in the work of dGe lugs writers like mKhas grub rje, and their approach—which would see the non-conceptual experience as in fact having a structure and therefore capable of being conceptualized, while not wishing to detract from the perceived nature of the experience itself—seems to me to be a necessary move towards greater systematic coherence. It is worth noting, however, that the very dGe lugs need for systematic coherence on such issues was itself a source of criticism by Tibetan scholars of other schools. I would argue however that this need was not religiously irrelevant or arbitrary.

To exist (*yod pa*), in dGe lugs thought, is to be a referential cognitive object (*shes bya*).[24] These two expressions refer to the same class. Thus if something is or can be a referential cognitive object, it exists. It follows from this, given the dGe lugs position on the ultimate as a cognitive object, that the ultimate too exists. Tsong kha pa himself comments that if the ultimate does not exist it could never be cognized, and the holy path would be pointless. Which is to say that we could not both cognize it *and* say that it could not be an object of cognition.[25] Tsong kha pa does not intend this as a *proof* of the ultimate. He is speaking to co-religionists. What he wishes to show (against those like Sa skya Paṇḍita) is that the ultimate must *exist*, that is, it must be capable of standing as a cognitive referent, and its non-conceptual nature does not entail that it cannot be conceptualized as, here, 'existent'. Even with reference to the non-conceptual, some concepts are not anathema, but can be applied and applied correctly. Since it exists a cognition of the ultimate must take a referential object.

In a famous comment the Indian Madhyamaka writer Candrakīrti (seventh century) remarks that whether Buddhas occur or not, the true nature of things remains. It exists (*chos nyid ces bya ba ni yod do*). It is the essence (*rang bzhin*) of things such as the eye and so on, by which Candrakīrti means it is the essence of all things. It is their very own nature which is to be directly cognized (i.e. is a referent of a direct gnosis) by an awareness which is free from the obscurations of nescience (*ma rig pa'i rab rib dang bral ba'i shes pas rtogs par bya ba'i rang gi ngo bo'o*).[26] Thus the ultimate is the essence of things, their true

[24] See Hopkins, 1983, pp. 214–5.
[25] Tsong kha pa's commentary to the *Madhyamakāvatāra*, the *dBu ma dgongs pa rab gsal*, p. 424: *don dam pa'i bden pa med na ni de rtogs pa mthar thug pa 'gro ba med la/de med na lam sgom pa don med par 'gyur te/*. For more on this and related themes see Williams, 1982.
[26] See Candrakīrti's *Madhyamakāvatārabhāṣya* on 6: 181–2. I have used the Cone edition, mDo xxiii, ff. 217–350.

nature, and it can be known as an intentional object in the mental state (which is therefore a mental *act*) of direct non-conceptual awareness. Tsong kha pa defines the ultimate truth as 'that which is found by a critical analytic inferential awareness which sees an intentional referent that is actually the case' (*yang dag pa'i don mthong ba'i rigs shes kyis rnyed pa don dam bden pa'i mtshan nyid du gsungs pa'i phyir ro*).[27] If we take this comment together with that of Candrakīrti, we can see that the very same ultimate can be known—albeit in different ways— directly through acquaintance in the non-conceptual gnosis, and through knowing that it is the case through analytic reasoning which must necessarily be conceptual. From which it follows, of course, that whatever is known in the non-conceptual gnosis cannot be at variance with what is known in analytic reasoning. The gnosis for Tsong kha pa must therefore have a content, it must be conceptualizable.

Candrakīrti referred to the 'essence of things such as the eye and so on'. By 'essence' he clearly means here their ultimate nature. And that ultimate nature is, for Madhyamaka writers like Candrakīrti, Tsong kha pa and mKhas grub rje (and Kamalaśīla too, although according to Tibetan doxographers he follows a different sub-school of Madhyamaka) in the words of mKhas grub rje 'their not-being-established ultimately'. He comments, using a pot as his example: 'One should know the following: The ultimate of the pot, the essence of the pot, and the final mode of being of the pot, is the not-being-established-ultimately of the pot.'[28] This is also called their 'emptiness' (Skt.: *śūnyatā*/Tib.: *stong pa nyid*). Thus the ultimate which we have been talking about is for these Buddhist thinkers a negation, not in the sense of a positive approached through a *via negativa*, but a simple negation. It is the very absence of ultimate existence, and it is taken to apply to all things including itself. For all x, if x exists (that is, can be a referential cognitive object), x is empty of ultimate existence.

Space prevents me from giving an extensive explanation of what is going on here. I have said more about it elsewhere.[29] The Madhyamaka offers a sustained critique of what Nicholas Rescher has called 'our "standard view" of natural reality, as a congeries of physical particulars emplaced in space and time and interacting causally'.[30] The perspective of Madhyamaka is that the world cannot be made up of inherently

[27] From Tsong kha pa's commentary to the *Madhyamakakārikā* of Nāgārjuna: *dBu ma rtsa ba'i tshig le'ur byas pa Shes rab ces bya ba'i rnam bshad Rigs pa'i rgya mtsho*, f. 237a.

[28] mKhas grub rje, 1972, p. 98: *bum pa don dam par ma grub pa de/bum pa'i don dam dang/bum pa'i rang bzhin dang bum pa'i gnas lugs yin no zhes shes par bya'o/*.

[29] See in particular Williams, 1989, pp. 60–72; and also Williams, 1982.

[30] See Rescher, 1973, p. 8.

independent particulars, and any claim to establish such a view of the world and events within it runs into paradoxes when subjected to critical examination. If we subject, say, a pot understood as an independent object out there in an independent external world to close critical investigation, we will find that such a pot cannot be. In Madhyamaka parlance, it is not found, lost, when subject to analytic investigation. And, as the eleventh-century Mādhyamika and important missionary to Tibet Atiśa puts it: 'If one examines with critical analysis this conventional (world) as it appears, nothing is found. The non-finding-ness is the ultimate; it is the primeval true nature of things.'[31] The ultimate is what is ultimately true about *all* things without exception, that they lack independent ultimacy. Although Nicholas Rescher's 'conceptual idealism' diverges at some, no doubt fundamental, points from the perspective of Madhyamaka, Candrakīrti, Tsong kha pa, Atiśa and others would have been happy with Rescher's observation that:

> Reality—our reality', as we can and do view it—is a 'mental construct' built up in the transaction of experiential encounter of person and environment by means of a conceptual framework that invariably and inevitably makes essential use of organizing principles. (1973, p. 4)

Madhyamaka writers do not see this as denying the existence of the world, but rather pointing out its actual status as dependently originated and lacking in the projections of immutability and mind-independence which we vest it with. *The* world is the only world; there is no Absolute at all—or rather the only Absolutes are the objects of projections of absolute nature which we mistakenly engage in *as if* they were really independent and self-subsisting entities. These projected Absolutes correspond to absolutely nothing.

Thus the ultimate in the Madhyamaka thought of dGe lugs writers like Tsong kha pa (or ultimates, for Madhyamaka frequently speaks of them in plural terms) is adequately represented in propositional form as the fact that *x* lacks ultimate—that is, inherent, mind independent—existence. Knowing it is a knowing *that*, and although it can be discovered through faith or scriptural utterance, i.e. the testimony of a reliable witness, this is not thought to be philosophically or in the last analysis religiously significant. It can and should be known through inference, critical thought, analysing the object of analysis to see if it does indeed have independent inherent existence. If this investigation

---

[31] *Satyadvayāvatāra* verse 21, in Lindtner, 1981: *kun rdzob ji ltar snang ba 'di | rigs pas brtags na 'ga' mi rnyed | ma rnyed pa nyid don dam yin | ye nas gnas pa'i chos nyid do |.*

is not carried out *by someone* then the ultimate—absence of inherent existence (*niḥsvabhāvatā*)—could not be known. If it is not carried out *by oneself*, then a knowledge based only on belief or hearsay would not have the force to uproot our habitual patterns of perception which vest things with inherent existence and therefore lead to corresponding patterns of egoistic grasping behaviour which fuel the process of unenlightenment.[32]

It should be clear therefore why Kamalaśīla felt the position of the monk Mahāyāna to be so mistaken, and accused it of being the very antithesis of a wisdom which results from a correct analysis of the way things are. It seemed obvious to Kamalaśīla that making the mind a blank could not lead to an appreciation of the ultimate, for the ultimate is something about things, and can only be discovered through investigating things themselves. mKhas grub rje comments that those who wish to attain enlightenment need first to determine what reality is like, i.e. how things really are. If they do not, but simply strive to enter a state of non-conceptual absorption, then this is of no final soteriological significance—it is not really even Buddhist (1972, p. 6). The root delusion which generates unenlightenment is self-grasping (*bdag 'dzin*). Therefore only by actively uprooting this grasping through

---

[32] In a letter to me after the original delivery of this paper Michael McGhee has raised the interesting objection here that it would be incorrect for the Buddhist to hold that we *believe* things to have inherent existence just because we do not realize that they are dependent on conditions and therefore impermanent, etc.: 'We may not have any thoughts on the matter. The fact that I do not realize that lightning is an electrical discharge does not entail that I believe that it's *not*.' This is a complex issue for the Mādhyamika Buddhist, who certainly does maintain, with McGhee, that it would be wrong to say in any *simple* way that 'the man in the street' should be taken to hold the truth of propositions which he has never entertained or articulated. In general I think the Mādhyamika might reply that it all depends what we mean by 'believe'. It is arguable that there are all kinds of things which I could be said to believe although I do not have any thoughts on the matter, evinced in my behaviour such that, if it were said that my behaviour entails an acceptance of the truth of proposition *x*, I would assent to the truth of proposition *x* or change my behaviour in a way that does not require such assent. For example, the Mādhyamika would urge that there is a sense in which we can be said to believe that things are not mind-dependent not because we have thoughts about the subject, but because an assent to the truth of the proposition 'Things are not mind-dependent' is implicit in our behaviour towards things. It forms part of a framework for a system of beliefs and perceptions which the Buddhist holds is radically skewed because the framework itself is skewed. We are touching here on the Madhyamaka treatment of *latent, innate* tendencies to misperception implicit in what it is to be unenlightened. Needless to say, it is far too large a topic for a footnote!

understanding emptiness of inherent existence could one be liberated (ibid. p. 7).

Knowing the ultimate is a knowing that, and is therefore, of course, through and through conceptual. Without concepts there could be no knowing the ultimate. Yet there are different levels of conceptual usage. dGe lugs writers do not see a genuine uprooting of the habitual patterns of infinite lifetimes springing solely from an intellectual conviction, even if that conviction were rooted in inferential discovery. Repeated analysis, repeatedly discovering the same fact, strengthens the conviction, but cannot uproot deep habitual patterns. If I find again and again that my reasoning shows Archibald to be a thief, I become more and more certain that this is indeed the case, and I start to behave in a different way towards Archibald. But if I see him actually stealing the suet pudding, does a *qualitative* change in my attitude to Archibald take place? Perhaps all lingering remnants of respect are lost. In the case of emptiness, while there is a knowing that, there is also said to be a mental state of unwavering direct acquaintance with the very emptiness itself, the very absence which is an absence of inherent existence. I confess this sounds a bit peculiar but, providing we are clear this experience is not to be identified with a mere blank mind, I am not sure we are in any position to deny its possibility, or the dramatic results expressed in terms of the Madhyamaka system which are said to follow from it. Sartre, of course, speaks of seeing the absence of the expected Pierre in the cafe as a direct seeing of a positive absence.[33] In dGe lugs thought too a non-entity, an absence, which includes an emptiness, is nevertheless an existent, since it can form the referential object of a cognitive act. We can perhaps make some sense of a *yogin* having the ability to focus unwaveringly on the very absence itself, and we can perhaps grant also that if he could do this perfectly, then since the object would be an absence alone he would be in a mental state lacking all other objective content. And I think we have no grounds for *denying* that this could begin to uproot even the habits of infinite lifetimes. Having said that, I confess a certain unease here. On dGe lugs terms this non-conceptual mental state has a subject—the *yogin*'s mind—and an object, emptiness. But emptiness is a mere negation. Can there be a mental state with *only* a negation as its object? How could one in experience distinguish this from no experience at all? Does it not in the end fall prone to the very criticisms we made of the monk Mahāyāna's blank mind? As we shall see, other schools than the dGe lugs aimed their criticisms precisely at what they saw as the inadequacy of this notion of emptiness to support the Buddhist path to final liberation. The liberating gnosis, they argued, must have a more positive content.

---

[33] See his discussion in Sartre, 1966, pp. 10ff.

Anyway, one thing should be quite clear at this point. Cognizing emptiness, the ultimate, and uprooting these habitual mental tendencies, is for any school only half of what is thought of as enlightenment in Madhyamaka. To have an absorption directed towards emptiness is only valuable precisely *because* it uproots egoistic grasping. In itself it is incomplete. The meditator does not remain in a state of emptiness-absorption, but integrates that awareness into everyday life so that he or she can operate in the world for the benefit of others, with no egoistic concern, but complete altruism.

Thus the goal here is not simply to cognize the ultimate. Indeed, even this knowledge of emptiness through acquaintance has stages to it, for dGe lugs texts speak of the absorption occurring through the medium of concepts, and then eventually in a direct non-conceptual absorption which nevertheless is the result of all that has gone before, and is in content terms in no sense at variance with it.[34] As non-conceptual this absorption involves no language, no (conscious) placing within classes. It is therefore seeing the very absence, emptiness, completely uniquely, in an experience which is held to be like water entering water. There is said to be no sense that 'Ah, this is emptiness!' How, therefore, can we know that it *is* an experience of emptiness, the same as was previously experienced conceptually. The answer, I think, has to come at least in part from the conceptual framework within which the meditation is occurring, the stages which led up to it, which entail that subsequently the content of the non-conceptual experience can be uncovered. From this it would follow that there could be no grounds for claiming that two non-conceptual experiences occurring within different theoretical and practical systems are of the same thing, for the processes which led to the experiences determine what they are of. Thus Kamalaśīla was surely right not only in condemning the monk Mahāyāna's blank mind for denying the processes which lead to the non-conceptual experience of emptiness, but also in implying that the non-conceptual experience referred to by the monk Mahāyāna could not be the genuine liberating non-conceptual experience of Madhyamaka.

The dGe lugs perspective on these matters was not the only one in Tibet. Traditionally and to the present day there is said to be three other schools of Tibetan Buddhism: rNying ma (pr.: Nying ma), bKa' brgyud (Ka gyer), and Sa skya (Sa kya), of which Sa skya Paṇḍita was a

---

[34] The highly sophisticated epistemological theories and techniques which dGe lugs writers brought into play in order to explain how a conceptual process can lead to a direct non-conceptual awareness could form the subject of a further very long paper! For some materials in English see Klein, 1986, especially ch. 9, and Lati Rinbochay, 1980.

particularly renowned exponent. The sixteenth century bKa' brgyud writer Dvags po bKra shis rnam rgyal (Dakpo Trashi Namgyel) refers to the mere absence which is discovered through Madhyamaka analysis as tantamount to nihilism, and implies that this is not the emptiness which his school takes as the true ultimate.[35] The controversial Sa skya scholar Shākya mchog ldan (1428–1507) appears to have maintained that the ultimate associated with Madhyamaka analysis is a destructive ultimate (*chad stong*), a mere negation (*med dgag*), which is a form of negation which does not imply any contrasting reality. He seems to have held that this teaching is valuable for realizing what does not exist, clearing the field, as it were, but at the time of meditative absorption aimed at the ultimate one has to leave behind such analyses, and such a perspective.[36] The 'mere negation' (*med dgag*/Skt.: *prasajyapratiṣedha*), which is a straight negative carrying with it in context no implication of a positive, is often contrasted in Indo-Tibetan thought with an implicative negation (*ma yin dgag*/Skt.: *paryudāsapratiṣedha*), which does eventually imply a positive in place of what is negated.[37] Thus in criticizing the Madhyamaka perspective of the dGe lugs, Shākya mchog ldan and others are saying that its ultimate is nothing more than a nothing. The dGe lugs would of course accept this. It is not just any old nothing, though. Rather it is the very absence of inherent existence—that, the false projection of which, has kept us in unenlightenment. For Shākya mchog ldan this mere absence is a deficiency—in reality there is a positive ultimate which is beyond all conceptual determinations and beyond the ultimate referred to in the dGe lugs, which is merely the ultimate truth about the phenomenal world. Correspondingly, the emptiness which is pure negation and is known through critical analytic reasoning is thereby just a conceptual emptiness. The true, highest, ultimate is known not through analysis (which would necessarily be conceptual), but placing the mind simply, unwaveringly, in non-conceptual absorption. This, to the dGe lugs, is simply the position of our old friend, the monk Mahāyāna.

[35] See Namgyal, 1986, p. 78. He takes his support here from a verse in the *Kālacakra Tantra*, attributed to the Buddha himself. Unfortunately I do not possess a copy of the Tibetan text used for this translation. I have my doubts about its complete reliability however, at least as regards the straight translation of Madhyamaka material. Compare bKra shis rnam rgyal's comments on the dGe lugs (Madhyamaka) emptiness not being the real ultimate with his contemporary the Eighth Karma pa, Mi bskyod rdo rje's (1507–54) attack on dGe lugs Madhyamaka in Williams, 1983, especially pp. 134 ff.

[36] See the highly critical account in the *Grub mtha' shel gyi me long*, by the dGe lugs scholar Thu'u bkvan bla ma bLo bzang chos kyi nyi ma (1737–1802), Beijing edition, 1989, p. 231. This has been translated in Ruegg, 1963.

[37] For the dGe lugs view of these negations, see Klein, 1986, chs. 6, 7.

Thus the dGe lugs are accused by its opponents of remaining in conceptualization. The dGe lugs accuses its opponents of denying the primacy of analysis and thereby in fact if not intention teaching no insight at all, but only a tranquil, stupified mind which is not truly Buddhist. Even as early as Rong zom Paṇḍita Chos kyi bzang po (pr.: Cher kyi zang po), a rNying ma writer of the late eleventh century, we find him answering the accusation that his teachings are opposed to logic and should not be accepted, with the reply that reasoning and logic apply mainly to conceptual thought, to sensory objects concerned with *saṃjñā* (*'du shes*). They do not apply to the 'essence' (*ngo bo nyid*), which is the object of a mind particularized by stainless wisdom.[38] This view is common in the rNying ma tradition of the Great Perfection (*rdzogs chen*), where it is taught that those of sharp intellect do not need to engage in analytic investigation at all. Indeed, they do not even need to meditate, for even meditation involves conceptual awareness.[39] I have argued that such a perspective entails a difference with the dGe lugs in what the ultimate is, and this appears to be accepted by writers like Shākya mchog ldan. One corollary of this, however, is that there appears to be a case against the common view stated nowadays among Tibetans that all the four systems are aiming at the same goal. I have also argued in defence of the dGe lugs that there is no immediate contradiction between claiming that the ultimate can be conceptualized, and it is known in an experience which does not appear to the one undergoing it to be conceptual.[40] The fact that emptiness is a

[38] *ngo bo nyid kyi mtshan nyid ni shes rab dri ma myed pas bye brag tu byas pa'i blo'i spyod yul lo/.* Text in Karmay, 1988, pp. 128–9.

[39] See the translations from the work of the greatest rNying ma scholar, kLong chen rab 'byams pa (Long chen rap jam pa: 1308–63), in Tulku Thondup Rinpoche, 1989, pp. 282–4. See also kLong chen pa's comments on the delusions of logicians, pp. 267–9.

[40] Of course, I have suggested that there may be 'subconscious concepts' involved. I am sure that Shākya mchog ldan would see this as a point to attack the dGe lugs. But the non-conceptual experience of the ultimate must be conceptual*izable*. This would only be a criticism if one could coherently argue that 'non-conceptual' here entails non-conceptualizable. But I think it cannot, and the systems of the schools show it cannot. Actually, the presence of subconscious concepts engaged in making sense even of the non-conceptual experience of emptiness may be one way for the dGe lugs to answer the problem of how to distinguish an experience which takes a mere negation alone as an object from no experience at all. There could, for example, be some sort of 'subconscious conceptual traces' still operative present from the previous analyses which led up to the non-conceptual experience. An alternative (or maybe complementary) approach would be to argue that what the dGe lugs tradition is actually trying to do is describe an experience which is in fact an experience of pure, radiant consciousness which is *as if* it had no intentional

non-entity does not for the dGe lugs mean that it cannot be experienced in a non-conceptual experience, nor does it mean *as such* that this experience could not liberate from unenlightenment. Moreover the non-conceptual ultimate of Shākya mchog ldan *et al*. would also seem to be open to criticisms of conceptualization as soon as it is mentioned, and it would be exposed to a dGe lugs Madhyamaka critique directed at all real, positive Absolute Realities. It is not enough to reply simply that, well, the attack would work were our Absolute conceptualizable, although it is not. That would be to try and have the suet pudding and eat it! I have suggested also that if knowing the ultimate in a non-conceptual experience involves non-conceptual*izability* then this is either incoherent or useless for Buddhism. And I have expressed some reservations about whether the dGe lugs perspective is nevertheless in

referent, although we know from the previous analyses that it has a referent—without the referent it would not be the pure, radiant consciousness we are talking about—and this referent is in fact the same emptiness discovered conceptually through analysis. This experience has been brought about through Madhyamaka analysis, as its culmination. Writers like Shākya mchog ldan, it would then be argued, are simply wrong in thinking that realizing a pure, radiant consciousness requires abandoning analysis, or seeing its limitations, postulating a real, inherently existing positive Absolute Reality in some sense identified with that radiant consciousness. I do not think this approach would be radically at variance with the dGe lugs tradition. Having said that, I am still not sure that I can make any sense of a mental state which is said to be pure consciousness without an object—or rather, with an object which is a mere absence. I can accept that it contains within it all that has gone before, and this distinguishes it from a blank mind. But I still cannot see that this pure consciousness with a negative object can be experientially distinguished from unconsciousness. In other words, I am not clear that we could speak of such a mental state of pure consiousness as an *experience*. I am doubtful that it makes sense to talk of contentless consciousness, and saying that it has content in the form of an intentional referent which is a mere absence alone does not seem to me to solve the difficulties. Something more is going to have to be said to make it an experience, it must have greater (conceptual) content. This something more may be related in some way to previous acts of analysis, i.e. what has gone before. Can the notion of 'subconscious conceptual traces' help? It depends on what they are supposed to be. But it might be suggested that what I am doing here is anyway pointless, since I am speculating about a paranormal experience which I confess I have not had. It should be clear however that what I am worried about is not the attempt to describe that which is held to be ineffable but whether, given what the experience is said to be, it makes any sense at all. In other words, the conceptual issue of whether there can be an experience *of that type*. For more on the perspective of Shākya mchog ldan (although not his own particular views) and the similar Jo nang tradition in Tibet, and also its origins within Mahāyāna Buddhism, see Williams, 1989, ch. 5. For a sympathetic account of this perspective see Tsultrim Gyamtso Rinpoche, 1986.

the last analysis coherent in maintaining that there can be a non-conceptual experience which takes a mere absence alone as its object. There are still questions in my mind about whether it makes sense to talk of an *experience* of this type.

One feels this controversy between the schools could run and run. It did, and I have! I shall stop here.

References

Demiéville, P. 1952. *Le Concile de Lhasa*. Bibliothèque de l'Institut des Hautes Etudes Chinoises 7. (Paris: Presses Universitaires de France).

Geach, P. 1971 reprint. *Mental Acts* (London: Routledge and Kegan Paul).

Heath, P. L. 1967. 'Concept', in *The Encyclopedia of Philosophy*, P. Edwards (ed.) (New York: Macmillan and Free Press).

Henle, P. 1970. 'Mysticism and semantics', in *Philosophy of Religion*, S. M. Cahn (ed.) (New York: Harper and Row).

Hopkins, J. 1983. *Meditation on Emptiness* (London: Wisdom Books).

Houston, G. W. 1980. *Sources for a History of the bSam yas Debate*. Monumenta Tibetica Historica 1:2. (Sankt Augustin: VGH Wissenschaftsverlag).

Jackson, D. P. 1987. *The Entrance Gate for the Wise* (Vienna: Wiener Studien zur Tibetologie und Buddhismuskunde).

Karmay, S. G. 1988. *The Great Perfection* (Leiden: E. J. Brill).

mKhas grub rje dGe legs dpal bzang. 1972. *sTong thun chen mo*, ed. Lha mkhar yongs dzin bstan pa rgyalmtshan. (New Delhi, Madhyamaka Texts Series 1).

Klein, A. 1986. *Knowledge and Liberation* (New York: Snow Lion).

Lati Rinbochay. 1980. *Mind in Tibetan Buddhism*, E. Napper (ed.) (London: Wisdom).

Lindtner, Chr. 1981. 'Atiśa's introduction to the two truths, and its sources', *Journal of Indian Philosophy* **8**, 161–214.

Mundle, C. W. K. 1970. *A Critique of Linguistic Philosophy* (Oxford: Oxford University Press).

Namgyal, T. T. 1986. *Mahāmudrā: The quintessence of Mind and Meditation*, trans L. P. Lhalungpa (Boston: Shambhala).

O'Flaherty, W. 1975. *Hindu Myths* (London: Penguin).

Rescher, N. 1973. *Conceptual Idealism* (Oxford: Basil Blackwell).

Ruegg, D. S. 1963. 'The Jo naṅ pas: A school of Buddhist ontologists according to the *Grub mtha' šel gyi me loṅ*', *Journal of the American Oriental Society* **83**, 73–91.

Sartre, J.-P. 1966. *Being and Nothingness*, trans. H. E. Barnes (New York: Washington Square).

Sastri, N. A. (ed.) 1938. *Bhavasaṅkrānti Sūtra and Nāgārjuna's Bhavasaṅkrānti Śāstra*. (Madras: Adyar Library).

Thu'u bkvan bla ma bLo bzang chos kyi nyi ma. 1989. *Grub mtha' shel gyi me long* (Beijing: Kun su'i mi rigs dpe skrun khang).

Tsong kha pa. 1966. *dBu ma rtsa ba'i tshig le'ur byas pa Shes rab ces bya ba'i rnam bshad Rigs pa'i rgya mtsho*. Modern blockprint made in India.

Tsong kha pa. 1973. *dBu ma dgongs pa rab gsal* (Sarnath: Pleasure of Elegant Sayings Printing Press).

Tsong kha pa. 1973. *Drang nges legs bshad snying po* (Sarnath: Pleasure of Elegant Sayings Printing Press).

Tsultrim Gyamtso Rinpoche. 1986. *Progressive Stages of Meditation on Emptiness*. trans. Shenpen Hookham (Oxford: Longchen Foundation).

Tulku Thondup Rinpoche. 1989. *Buddha Mind* (New York: Snow Lion).

Vaidya, P. L. (ed.) 1960. *Bodhicaryāvatāra* (Darbhanga: Mithila Institute).

Von Stael-Holstein, A. (ed.) 1926. *The Kāśyapaparivarta* (Shanghai: The Commercial Press).

Williams, P. 1980. 'Some aspects of language and construction in the Madhyamaka', *Journal of Indian Philosophy* **8**, 1–45.

Williams, P. 1982. 'Silence and truth—some aspects of the Madhyamaka philosophy in Tibet', *The Tibet Journal* **7**, 1–2, 81–90.

Williams, P. 1983. 'A note on some aspects of Mi bskyod rdo rje's critique of dGe lugs pa Madhyamaka', *Journal of Indian Philosophy* **11**, 125–45.

Williams, P. 1989. *Mahāyāna Buddhism: The doctrinal foundations* (London: Routledge).

Williams, P. 1991. 'Some dimensions of the recent work of Raimundo Panikkar—a Buddhist perspective', *Religious Studies* **27**, 511–21.

'Know Thyself':
What Kind of an Injunction?

ROWAN WILLIAMS

To be told, 'know thyself' is to be told that I don't know myself *yet*: it carries the assumption that I am in some sense distracted from what or who I actually am, that I am in error or at least ignorance about myself. It thus further suggests that my habitual stresses, confusions and frustrations are substantially the result of failure or inability to see what is most profoundly true of me: the complex character of my injuries or traumas, the distinctive potential given me by my history and temperament. I conceal my true feelings from my knowing self; I am content to accept the ways in which other people define me, and so fail to 'take my own authority' and decide for myself who or what I shall be. The therapy-orientated culture of the North Atlantic world in the past couple of decades has increasingly taken this picture as foundational, looking to 'self-discovery' or 'self-realization' as the precondition of moral and mental welfare. And the sense of individual alienation from a true and authoritative selfhood mirrors the political struggle for the right of hitherto disadvantaged groups, especially non-white and non-male, to establish their own self-definition. The rhetoric of discovering a true but buried identity spreads over both private and political spheres. The slogan of the earliest generation of articulate feminists, 'The personal is the political', expresses the recognition of how this connection might be made.

R. D. Laing's seminal work of 1960, *The Divided Self*, did much to popularize the idea of a distinction between different 'self-systems', with the essential feature of schizoid disorder being defined as the separation of a 'real', 'inner' self, invisible to the observer, from the behaviour of the empirical ('false') self. For Laing, the clinical schizophrenic's condition is an extreme case of the schizoid fantasies common in supposedly sane persons, whose behaviour and language betray a belief that they have an untouched core of selfhood which must not be compromised or limited by involved action, but which lives in a state of fictitious freedom and omnipotence—described by Laing (pp. 87–8) as the direct opposite of Hegel's insistence in the *Phenomenology* that performance alone measures what is real in the life of an agent. Laing, in fact, is diagnosing the language of a 'real', non-appearing self as a sign of dysfunction; but already in *The Divided Self* and more dramatically in some of his later writings, he is also suggesting that the dysfunction is virtually forced on vast numbers of persons because the public realm of language and action is systematically oppressive and distorting. From this aspect of Laing's thought, reinforced by his abundant use of Kafka and Sartre, it is not difficult to slip into the view that the socially-constructed and socially-sustained self is indeed false in some absolute sense, and that authenticity lies in a hidden dimension, a core of uncompromised interiority. This is a conclusion which Laing himself is very careful not to draw; but a superficial reading, aided by existential-ist and oriental ideas (imperfectly digested), could produce the para-doxical doctrine that the 'true' self is present but inoperative, and may be discovered by bracketing out large tracts of the social, the corporate, the linguistic. This is in some respects obviously the child of the classical project of psychoanalysis, the decoding of present linguistic and symbolic behaviour so as to uncover the conflicts which generate my current self-presentation. But the important difference is that, for the searcher for the lost, 'true' self, the business of penetrating behind self-presentation leads *beyond* buried conflict to an authoritative source or centre of energy. Self-knowledge thus becomes more than an acquaintance with the history of trauma and defence, and appears as the possibility of liberating contact with a power that can transform present performance, replacing a false system of self-representation with another system which does not systematically mask real desires and needs. This may be conceived, picking up the clues of eastern religious philosophy, as *the* Self, the divine undifferentiated reality within; or as an individual system of immanent forces in balance, a temperamental pattern of gifts, characteristic affective responses, undistorted desires. It is, in either case, habitually pictured as present but concealed. The archaeology of analysis reveals, eventually, a living subject with an agenda distinct from what has been the agenda of the habitual self-awareness. From one point of view, this scheme represents a quite remarkable rearguard action fought by romanticism against the dissolu-tion of the autonomous agent threatened by analytic disciplines—remarkable because it leads analysis inexorably back towards a pre-Freudian mythology of unambiguous nature, the naked self, prior to history and conversation. The philosophical problems of this naked self are tediously familiar. In this particular context, we should have to ask: how can a present discourse, shaped by the history of my speaking and hearing, intelligibly claim to re-establish what is not so shaped? how could such a claim be tested? what sense can we make of the idea of a self with specific dispositions and desires prior to relationships if the self is self-aware only against the presence of a resisting or interrupting other? The archaeological analogy is question-begging (how does the com-promised active self recognize what it discovers as its own truer reality?): the 'discovered' self is surely the construct of present actions and interactions—an insight authoritatively mapped by Lacan in all its complex ramifications.[1] For Lacan, self-knowledge is precisely a recognition of the dialectical nature of being a 'subject', the inescapable involvement of the self in the desire of the other: it is, in the analytical encounter of the present moment (in which the self is confronted by another self—the analyst—trained, as far as possible, to set aside the ego-system obsessed with the meeting of desires) the recognition that the ego conceals the way in which the subject fundamentally exists—as lack, as the wanting of what it is not, and thus both relational and self-cancelling (pp. 11, 31, 62–3, 83–5). If there is a secret to be uncovered, it is that there is nothing prior to reciprocity. The way in which the ego habitually organizes itself is as a system potentially or really *in possession* of what it desires: but the subject is constituted as what does not possess, desiring and desiring to be desired. Self-knowledge delivers not a hidden, authentic agenda to replace our current troubled system of self-representation, but a sense of one's irreversible engagement in an exchange with no substantial fixed points: it locates us more firmly now in the complexities of exchange and teaches us not what we must do to be true to our 'nature' but simply to be endlessly iconoclastic about the claims of the ego. Self-knowledge amounts to a practice of conversational self-questioning, and the 'true' self is no substance but simply the enacting of such a practice. The ego, of course, is never simply removed or dissolved: its formation is a primordial and (again) irreversible misunderstanding of what the subject is, and so it is what makes self-questioning both possible and necessary. I am indeed alienated from truth, and the substantive 'I' is the sign of that recognition; but this does not mean an apotheosis of instinctual existence or an attempt to dissolve historical consciousness (despite the vitriolic criticisms of Ferry and Renaut (1990, ch. 6) on this point).[2] Conscious life is what is set up by the tension between subject and ego: truthful consciousness acknowledges this and understands that the subject's presence in history and language depends on the false concreteness of the ego (demands and gratifications, goals and fulfilments), *and* that these fixities are constantly being subverted, *and* that the very *movement* of speech depends in turn on this subversion.

Lacan is worth pondering because he presents almost an inversion of romanticism and the Sartrean pathos of the frustrated true self. The

[1] I rely chiefly on Lacan's *The Language of the Self. The Function of Language in Psychoanalysis*, trans. with notes and commentary by Anthony Wilden (Baltimore, 1968).

[2] Their treatment more or less ignores the Lacanian emphasis on the analyst as catalyst of proper intersubjectivity, though it does identify elements of primitivist romantic pathos in some of Lacan's discussions (pp. 197, 203).

hidden and uncorrupted subjectivity which is somehow present as a realm into which I can escape is the most fundamental of all misunderstandings because there is no desire which is not already mediated—i.e. in some sense alienated. My 'I' is given, learned from the other; beyond it stands not a coherent and unified selfhood but—for Lacan—something like a foundational absence, a state of death. The subject's quest for itself is for him a desire for death. Yet this shocking recognition enables the recognition that the 'satisfaction' of the subject is not after all intrinsically at odds with the satisfaction of all subjects. The analytical conversation lays bare the fraudulence of the ego; the analyst's minimal ego enables the analysand's ego to be relativized, and intersubjectivity to appear, the reciprocal recognition of subjects (Lacan, pp. 84–5). The unclarities and points of strain in Lacan have been amply discussed, and I have no intention of simply presenting his account as incontrovertible truth (I find the status of the subject as primordial absence and Lacan's thesis about the subject's fulfilment in death especially problematic: Girard's critique (1987, pp. 403–5)[3] of Lacan on this point is pertinent, arguing that Lacan is still not free from the mythology of pre-cultural desire and primary self-constitution). But in so far as Lacan offers a uniquely full and acute critique of the Hegelian 'noble soul' as the terminus of self-knowledge, he makes it clear that, if self-knowledge is liberative, it is not because it issues in an authoritatively self-defining subject. The point applies politically as well as psychologically, and must stand as a question to (for example) essentialist and archaeological discourse within feminism. But the purpose of this essay is to look at the rhetoric of self-knowledge in the religious context; and since the language of 'true selfhood' and certain techniques of self-examination and self-appraisal, loosely grounded in psychoanalytical theory, are enjoying extremely wide currency in literature about Christian spirituality,[4] it seems as well to begin with some general

[3] 'Lacan falls into the error that is shared by the whole psychoanalytic school when he writes about capture by the imaginary—a desire that is not inscribed within the system of cultural differences and so could not be a desire for difference, but necessarily bears on something like the same, the identical, the image of one's ego, etc.' (p. 404). The point is a crucial one: the Lacanian view of the subject as directed towards death is only intelligible if there is indeed some kind of reality prior to being spoken to, being engaged with, even if this is conceived only as a notional or regulative level of the psyche's life. Even in such a minimal form, it affirms a priority of sameness over difference and nature over culture, which it is Girard's aim (in his own terms) to demythologize.

[4] I have in mind especially the varieties of popularized Jungianism now vastly influential in books and courses on spiritual direction—the Myers–Briggs personality typology, the categories of the 'Enneagram', and so forth. These techniques of analysis have great practical usefulness, as many experi-

reminders of the current difficulties in discussion self-knowledge. What I propose to do in the rest of this paper is to look briefly at three ways in which the injunction to self-knowledge has been used in Christian tradition, so as to pose two questions: first, are the traditional usages vulnerable to the critique of a post-Lacanian (and post-Wittgensteinian) account of the self? and second, does the contemporary Christian interest in self-knowledge belong in the same frame of reference as the language of earlier writers? My tentative conclusion will be that the answer to both these questions is 'No', and that some aspects of earlier Christian language about self-knowledge leave open the possibility of a useful conversation with the recent discussion I've mentioned.

My first example is the rhetoric of self-knowledge and self-recognition in the Christian gnostic literature of the second and third Christian centuries, in particular some of the texts from the Nag Hammadi collection. Fundamental to the mythology of all groups using gnostic idiom is the belief that our present human condition is enslaved by forgetfulness of our origin. Thus the *Apocryphon of John* (Robinson, 1977, pp. 99–116) describes how the ignorant world creator, himself oblivious of his origins, is tricked into imparting some element of divine spirit to the primordial human subject, who thus excites the jealousy of the cosmic powers who imprison this subject in matter and mortality, 'the bond of forgetfulness'. Adam is placed in Eden and told to eat and drink and enjoy himself (pp. 109–10). Divine grace hides in Adam the saving element of *epinoia*, intellectual grasp, but this has to be activated by a saviour who is first and foremost, 'remembrance' (p. 115). The *Gospel of Truth* (Robinson, pp. 37–49) accordingly describes the one who is saved from the wreckage of the cosmos as one who 'knows where he comes from and where he is going to' (p. 40). The sayings of Jesus in the *Gospel of Thomas* (Robinson, pp. 118–30) echo this frequently, but strikingly turn on the idea that the 'hidden' truth of who we are is in fact plain and obvious. 'If those who lead you say to you, "See the Kingdom is in the sky, then the birds of the sky will precede you . . . The Kingdom is inside of you, and it is outside of you. When you come to know yourselves, then you will become known . . . Recognize what is in your sight"' (p. 18); 'What you look forward to has already come, but you do not recognize it' (p. 123); 'The Kingdom of the Father is spread out upon the earth, and men do not see it' (p. 130).

---

enced spiritual directors confirm; but presented in terms of theory, they have some very questionable elements, not least in the language sometimes explicitly used of a 'purity of essence' preceding socialization, and in the mechanical and fixed ways in which personality types are sometimes presented in the self-help books generated by the popularity of this style of interpretation.

Examples could be multiplied, but the sense conveyed is clear. As in the Greek mysteries in which the original *gnōthi sauton* had its setting,[5] I must recognize what *kind* of a being I fundamentally am: self-knowledge here is nothing to do with individual self-analysis, a particular history, but is the discovery of the history I share with all the children of light—a pre-history, rather, in which all that constitutes particular history is the creation of a cosmic *ressentiment*. I have *lost* nothing in being subjected to the indignity of incarnation, but I am separated from my true self and my true home by forces external to myself who have 'clothed' me in a false identity (hence the significance in *Thomas* of the metaphor of stripping, as in saying 37, p. 122). *Thomas* presents the most sophisticated treatment of the theme, in suggesting that enlightenment is not simply knowing the myth of your origins and destiny, but understanding that the state of the world before your eyes tells you the truth of your nature: you are the seer, not the seen, and the material world lies before you, passive, like a corpse (sayings 56 and 80, pp. 124, 127). Understand that you understand, that you are not bound by reaction to what is before you, and you understand that you occupy a place beyond all worldly schemes of differentiation. You are the light in which things are seen. Where is the place where Jesus stands? it is the place from which light emanates, the interior of the enlightened person (saying 24, p. 121).

What we must recognize, then, is that our 'real' life is undifferentiated, a sort of Aristotelean self-noesis, indistinguishable perhaps from the life of what *Thomas* calls 'the All'. Gnostic self-recognition certainly deploys a rhetoric that has affinities with the romantic model of the buried, authentic self; but the buried self here is not the touchstone of authentic desire or unillusioned action, merely the purity of intellectual self-presence. In one way—unsurprisingly?—this veers towards the Buddhist ideal of self-knowledge, the dismantling of all specific desires, properties or projects so as to perceive the underlying absence of anything but the sheer 'thereness' of the empty abundance that is both blissful fulfilment and contentless void (*nirvāṇa* and *śūnyatā*). At another level, the analogies with a Lacanian analysis are striking, especially in the virtual dismissal of habitual bodily and emotional self-presentation as the effect of 'captivity' or falsehood, and the ultimate identification of the subject as absence. The difference, though, is at least twofold: the gnostic subject may be an absence, but could not be called a *lack*; it is beyond desire, and so beyond history. And the intersubjectivity that can be created in the analytic encounter is no part of the gnostic hope, which looks to a recovery of non-differentiation, a

[5] According to classical sources, it was inscribed at the door of the shrine of Apollo's oracle at Delphi.

divine sameness. From yet another perspective, however, gnostic rhetoric is indeed the distant parent of romantic mythology, in constructing not simply a picture of a world of error or misprision, but an actively hostile environment deliberately stifling the truth out of envy. The false and forgetful self is in no sense (as it would be for the Buddhist or the Lacanian) the formation, the *responsibility*, of the subject: it is wholly the creation of a hostile power. The forgotten self acquires the pathos of a victim—a move which decisively politicizes the language of gnosticism, giving the self a project of struggle against not alienated but nakedly alien force (though, once again, a document like *Thomas* shows how this language in turn can be demythologized and freed from the crude rhetoric of struggle against something external).

Gnosticism's difficulty is always in this last aspect of its schema. If history and the body are indeed *radically* alien to the spirit, the spirit is a stranger to guilt and division: it is not self-alienated but forcibly disguised from itself. Whence then comes the division? The problem is pushed back into the realm of the divine life itself, and there the same problems recur. Divine self-alienation (myths of the fall of Sophia) is no easier to theorize, and perhaps the only consistent solution to such a problematic was the absolute dualism of the Manichees. The rhetoric of gnostic self-knowledge represents a drastic working-through of certain features in the earliest Christian language we can trace—the summons to see one's ordinary self-presentation as deceitful (as in the synoptic Jesus' sharp antitheses between what is visible and what is 'in the heart'—Mt 5.21ff.— or the Johannine Jesus' allusions to the blindness of those who claim they can see—Jn 9.39–41—or Paul's regular opposition of 'flesh' and 'spirit' as moral systems); what it shows, though, certainly to the majority of writers in what was to become the Christian mainstream, is that to render the problem of self-deceit in such a way as to disclaim ownership of one's own deceptive history is to set up a problem as serious for metaphysics as for individual psychology.[6] The isolation of the non-responsible, inactive and impassive self finally requires the postulating, at one and the same time, of an inactive and impassive God and a universe of uncontrollably delusory systems.

Thus the relentless elaboration of the theme of self-deceit alone leads to the vision of a reality eternally and irreparably—and unintelligibly—

[6] Augustine in *Confessions* VII.2 summarizes the argument of his friend Nebridius about the difficulties of a thoroughgoing dualist metaphysic: either God is vulnerable to change and chance—in which case, it is perfectly conceivable that good will be defeated in the universe, and that therefore the good is not identical with the real; or God and the good are not vulnerable, in which case there is no need for a properly dualist theory in the first place. Divine self-alienation and primordial conflict between equipollent powers are equally insupportable positions in any intelligible metaphysic.

split, and to something like a technically schizoid construction of the world of the subject. A truth that is strictly incommunicable in habitual (bodily/temporal) self-presentation is liberative only negatively, as a relativizing of *all* determinations: it does not modify our negotiating of *particular* determinations. Consequently, as Christian language develops, the idea of an independent spiritual core to the person, a self untouched by time and guilt, recedes further and further: even with the pre-existent *nous* of Origen's anthropology, the present state of the self is intimately connected with the free self-determination of this primordial subject, which both is and is not involved in time and matter. The destiny of the spiritual subject is liberation, but, to arrive there, I must learn virtue *in* the school of the fallen soul and the empirical body. The *nous* is in principle independent of the body, yet it can only become what it should be—a spirit uninterruptedly contemplating God—through the life of material and temporal selfhood.

The difference between the language of gnostic ethics and spirituality and what finally became the normative idiom of Catholic Christianity is very clear if we turn to a second brief case-study, the discussion of self-knowledge in St Bernard's homilies on the Song of Songs.[7] Sermons 35 to 37 in particular deal with a text from the first chapter of the Canticle in the Vulgate, *si ignoras te . . . egredere* (Song of Songs 1.7): the Bride is told that if she fails to know herself, she must leave the sweets of contemplation for the exile of a life dominated by gratifying the needs of the senses. This, of course, is what the fate of Adam was; and why does Adam fall? By misunderstanding the meaning of his human dignity (35.6). He forgets that he is a creature (and thus dependent on God), and is expelled from Eden, thus becoming subject to a second and more serious ignorance of himself, forgetfulness of his rational and spiritual nature (35.7). Not knowing oneself, then, and not knowing God are intimately connected: if I do not know that I am God's creature, because I am hypnotized by the grandeur of the intellectual gifts given me, I shall not in fact know how to exercise those gifts, and I shall cease to be a rational creature at all (this rests on Bernard's general belief that, since reason in us is God's image, it cannot function when it does not have God for its object). Faulty self-knowledge has thus led to our present sad plight: what, then, does it mean to know myself truthfully now? It is to see my helplessness and loss, to discover that I now live 'in a region where likeness to God has been forfeited' (*regio dissimilitudinis*, 36.5). Yet simultaneously I must know that God continues to hear me and to give me grace: if I see myself as a fallen sinner, I must also see myself as a graced sinner, since

---

[7] Translated by Kilian Walsh, OCSO, in the *Cistercian Fathers* series (Kalamazoo, 1971 and 1976).

I could not truthfully know my sin without knowing God (I couldn't know what it was like to lose the image of God if I had no awareness of God). 'Your self-knowledge will be a step to the knowledge of God; he will become visible to you according as his image is being renewed within you' (36.6). Self-knowledge thus becomes the condition for repentance, prayer and practical charity (37.2): the Spirit of God begins to realize in us the dignity of God's children by forming us in holiness. So by our penitent recognition of what we are, we 'sow in tears'; but God's mercy guarantees that we 'reap in joy' (37.4, quoting Ps 125.5).

This is quite a complex depiction of self-knowledge. If we had to identify its focal themes, we could say that they are (i) the need throughout for a recognition that we are constituted 'rational' or 'spiritual' or whatever in virtue of relation to the creator, not as self-sufficient individualities, and (ii) the priority here and now of recognizing fallibility and failure as the self's truth, while perceiving also that such a judgment presupposes relation to God even in the acknowledgment that no proper relation yet exists. Thus there is no selfhood prior to the address or gift of God: reason responds to this, rather than having a simple primacy or autonomy. Even as a Godless and forgetful sinner, I am called into being as a self by the prior love of God. Self-knowledge makes no sense except as achieved in the face of God, in the light of God, and truthful self-knowledge establishes itself as such by incorporating recognition of the divine love; this is as true for Adam in paradise as for us now. The 'authentic' self is what I acknowledge as already, non-negotiably, caught up in continuing encounter with or response to divine action; and the acknowledgment is inseparable from converted behaviour. The person who knows him- or herself is manifest as such in the practice of prayer and almsgiving. Or, in short, the meaning of self-knowledge here is displayed in the performing of acts intelligible as the acts of a finite being responding to an initiative of generosity from beyond itself, an initiative wholly unconditioned by any past history on the self's part of oblivion or betrayal.

This is a model very nearly at the opposite pole from that of gnostic language. What to the gnostic is the terminus of the search for a true identity is here, in effect, the most lethal of errors: we do not arrive at a subject functionally identical with the worldless divinity, essentially indeterminate, but at a point of primitive determination, an irreducible status as hearer and recipient. The capacity to make sense of the world follows not from identity with the light of reason or order beyond the contingent present, but from grasping one's own contingency, articulating dependence (reason as *imago dei*). Identifying the substance of the self with a non-relational intellectual power is the pride which exiles us from Eden, and constantly pulls us to the sub-rational level of serving pure appetite (which we may successfully disguise as reasoning). Furthermore, our condition is historical, in the sense that, since there is no essence prior to relation with God to which we can have recourse, we are always what we have made of ourselves in encounter with God—not imprisoned angels, but struggling and inept conversationalists, whose errors and self-delusions build themselves into a formidable carapace of unreality, reinforced by every fresh stage in our self-representation unless interrupted by awareness of God—which is necessarily a silencing of our self-projecting. Thus there can be no authentic *image* of the self that has definition and fixity of itself (which would be another projection of the falsity of Adam's originary misperception)—there is the recognition of a history of error and failure to respond to what I now see or hear afresh (God), and the adoption of the kind of practice that militates against error about my metaphysical status—prayer and charity. There can be no *ressentiment* against the cosmos, since the imprisoning illusion is self-generated—reason's attempt to reflect on itself without the mediation of God's creating love. If there is a 'politics' to this account of self-knowing, it is not based on a rhetoric of reclaiming what has been taken away, or identifying a guiltless and uninvolved victim at the centre of my identity, but is orientated rather towards the suspicion of claims for the finality of self-definition, reasoned defences of the pursuit of interest and appetite on the part of a pseudo-self (individual or collective?) denying its contingency. The mistaken subject here is constructed as possessing, not lacking, power in the world of negotiation, and needing to recover a fundamental perception of *limited* power rather than a primordial liberty. That there is an ambivalence to this also is evident; more of that later.

Bernard's theology is decisively shaped by the heritage of Augustinianism, and it is to Augustine himself that we turn for a third and final perspective on the language of self-knowledge. The *Confessions* is a work permeated on practically every page by the acknowledgment that true knowledge of self is inseparable from true knowledge of God. *Cognoscam te Domine cognitor meus*, writes Augustine at the beginning of Book X, and, a little later (X.5), *quod de me scio, te mihi lucente scio.* As Book VII in particular makes clear, the recognition of God's absolute transcendence is a crucial moment in liberating Augustine from a crude and sterile picture of the self and its moral world: the problem of evil is reconceived, as a problem of the variability of the will in a contingent environment, rather than a question of how a substantial alien force could intrude itself into the world's fixed territory; the mind's evaluating and connection-making activities are seen as intelligible only if they take for granted an independent (non-worldly) measure of value and coherence. It is the utter and irreducible difference between God and the human mind that frees the mind to recognize itself and to be itself—to know that it is not a sort of material object, but is simply the activity of making sense of the world and of its own history: memory, says Augustine (X. 17), *animus est et hoc ego ipse sum*. I 'am' the recollecting and ordering of my past.[8] This is, by definition, an endless labour, and it can be carried forward only in the belief that there exists a full and just perspective on my history, not dependent on my own fallible perception. Truthful self-knowledge thus entails a constantly self-critical autobiographical project, striving to construct the narrative least unfaithful to the divine perspective. It will, of course, never *be* the divine perspective, because what God *sees*, I *learn* (and constantly, with every new action, must relearn). *Confessions* X describes vividly how the present awareness of what distorts judgment and desire enforces humility and a certain provisionality in our accounts of ourselves: what we can be certain of is not our own perseverance but the mercy of God, who alone sees what we are and what we need.

Some of these themes recur in the masterwork of Augustine's maturity, the *de trinitate*, whose eighth to tenth books are very largely devoted to a complex and subtle discussion of self-knowing. A full treatment would be difficult in the space available here,[9] but the salient points are these. Self-knowledge and self-love are brought into close connection, because of the recognition that the self is in motion to the Good (or at least what it thinks is the Good), that it cannot avoid making judgments of approbation and disapprobation: the self operates as if it knew more or less what it *wanted to be like* (*de trin.* VIII.iii–vi). We love good people, and part of what we love in them is their own love, their will that goodness be accessible to all (VIII.viii). So to know ourselves is to recognize our involvement in moral process by way of desire: we want to be just and loving, we learn justice and love from the just love of good people, and so we must understand ourselves to be so constituted as to be in love with loving (and so, ultimately, with the unreserved generosity of God). Book IX spells out some implications: we can't love ourselves without knowing ourselves, but we can't know without desire (IX.ii–vii). Consequently, in Book X, the paradox is stated in full force: 'total' self-knowledge is precisely the knowledge of

---

[8] Cf. R. Williams (1982), p. 29: 'The self *is* . . . what the past is doing now.' This formulation was criticized as fanciful and imprecise, but I should still want to defend it in so far as it represents the sense to which Augustine witnesses that present 'selfhood' is not an arena of open choices confronting an abstractly free volition, but a territory marked out by preceding determinations (by self and others), which mould, in ways frequently inaccessible to us, what can be and is done.

[9] I have attempted a slightly more extended discussion in Williams (1992).

the self as incomplete, as seeking (X.iii). Because it makes no sense to split the mind into the bit that knows and the bit that is known, then if I know myself as questioning and incomplete, as wanting to know, I know all I can know of what I am (X.iv). I know, very importantly, that I am neither God nor beast: a creature, but a reasoning creature (X.v: here is the theme of self-knowledge as knowing one's place in the order of things, which is found elsewhere in Augustine, and is obviously seminal for Bernard). I know too that my mind cannot be a material object, cannot be comparable to one of the things I think about (X.ix–x), because knowing the mind is always knowing from within a fluid activity, not a fixed external object.

I have tried to argue elsewhere[10] that Augustine's account here of the indubitability of self-knowledge may be closer to Wittgenstein's *On Certainty* than to Descartes, since it is above all an attempt to show that knowing the self is something quite distinct from any processes of 'coming to know' in which questions of evidence and relative degree of certainty can properly be raised (you couldn't be *fairly* sure you knew yourself on Augustine's account). Its most significant contribution to our present discussion, though, is the insistence upon the *unfinishable* nature of self-knowledge (a theme that can be paralleled, incidentally, in the Christian East, particularly in Gregory of Nyssa).[11] The self is in construction; the relating of a history is not the fixing of the self's definition or the uncovering of a hidden truth, but part of the process of construction, a holding operation. Furthermore, the process is bound up with the desire for the Good, for *iustitia*: the self in construction is a self whose good is understood in terms of a universally shareable good, and the self is not known adequately without a grasp of the inseparability of its good from the good of all. If there is a 'secret' to be uncovered by the search for self-knowledge, it is perhaps this unconscious involvement in desire for the common good; and if there is a 'politics' of self-knowledge in Augustine, it lies in the dissolution of any fantasy that the good can be definitively possessed in history by any individual or any determinate group in isolation. The distance between God as *summum bonum* and all creatures means both that there can be no settled state of absolute good for this world, and that there is

[10] Williams, 1992; for a perspective on the Wittgensteinian approach to self-knowledge, there is a useful article by Godfrey Vesey (1991).

[11] See the celebrated section in Gregory of Nyssa's *Contra Eunomium* II.107ff. (Migne, *Patrologia Graeca* 45,945Dff.; in the more recent edition of Jaeger, which corrects the numbering of the books in Migne, vol. I, pp. 258ff.) on our inability to give an account of what our own souls actually are in any finished way—a point designed to reinforce Gregory's insistence that the knowledge of God, in whose image the soul is made, is similarly unfinishable, eternally open or expanding.

(effectively) unlimited time for the working and reworking of corporate movement towards the Good.[12]

It is time now to attempt some drawing together of the threads of this diffuse discussion. I have suggested that the rejection of gnostic language about the hidden self left 'mainstream' Christianity with the task of dealing with its own fundamental vocabulary of self-deception and restoration without endorsing the mythology of a supra-historical subject, a self prior to and untouched by a history of interrelation and of determination by that interrelatedness. Bernard and Augustine present us with a self constructed in and only in contingency, and intelligible only as responding to address from beyond itself, never self-creating. For Bernard, if we do not see our rationality as tending Godwards, we become sub-rational: we need a de-mystifying of our intellectual and spiritual powers and an acquaintance with our powerlessness to avoid error, left to ourselves. For Augustine, we need to come to terms fully with our finitude and to recognize that our rationality is always emeshed in desire—most primitively, that desire for the good or the just that is obscured by our habitual misidentification of what we want, which results from the fictions of rivalry that corrupt the common life of human beings and reinforce an image of the self as an atomistic subject orientated to an endless series of specific gratifications (a consumer, in fact). For both Bernard and Augustine, the inaccessibility of the divine perspective is paradoxically liberating: there is always a resource for the renewal or conversion or enlargement of myself independent of what may happen to be my resources at any given moment, and there is always the possibility of more adequately ordering the telling of my life as I draw towards a perspective on myself undistorted by my self-interest—a perspective never possessed, never simply mine, but imaginable as a horizon against which other perspectives may be tested. And none of this would be conceivable if God were the occupier of a 'point of view' comparable to my own, a positional perspective like that of an ordinary subject, only larger.

There may, then, be a convergence of sorts with the picture outlined at the beginning of this paper. Self-knowledge is a practice of criticism, specifically the criticism of the way the subject distorts its self-perception into fixity by fixation upon the meeting of needs in the determinate

[12] 'Unlimited' time not, of course, in the sense that Augustine did not believe in human mortality or the end of the world at a determinate point; but in the sense that no term is fixed to the work of individual or society in the attainment of the good *within* history. We could never claim to have reached a plateau, nor is the failure to realize the fulness of God's justice within history an irreparable or unforgivable delinquency. All human achievement is provisional, all is therefore capable of flux for better or worse; there is, from *our* point of view in history, 'always' a future.

form in which they are mediated to us in the perception of the Other; and the ego-less interlocutor whose non-intervening presence exposes to us the possibility of this critical practice is identified as one who holds not even a residual position in the world where desires are negotiated. Theology's query to the Lacanian analytic project might be whether anything short of this horizon (of the essential absence from the world of the liberating interlocutor) enables us fully to see the possibility of a state of non-rivalry among human claims for satisfaction (Girard, Bk II, Bk III, ch. 1). Theology also assumes, however, that the subject's unthematized desire to return to itself is not a desire for a 'foundational absence', for death, but for a mindful standing in its basic position of creatureliness (unfoundedness in oneself), a standing before what always precedes it: neither the birth orifice nor the phallus, as in Lacan's orthodox Freudian idiom, but the intelligible Word that precedes even biology. Lacan's own regular advertence to the religious repertoire of images, Christian, Jewish and oriental, suggests that this parting of the ways is not a simple matter of one party taking a rationalist option for the clarities of natural science. There is more to be said from both sides.

I have hinted at a further convergence. Lacanian analysis goes a long way to removing the pathos of the victim from the destiny of the self, and Bernard and Augustine concur. But does this mean that the injunction to self-knowledge is primarily addressed to those who must be dispossessed? Both feminist and black theologies have often interpreted the summons to repentance, provisionality, the unmasking of pride, as inappropriate as addressed to them, as ideological commendations of passivity in an intolerable situation. The point is significant. We noted earlier the ambivalence of Bernard, if read as the language of the powerful (the male clerical ideologue) to the powerless—though we should remember that he is addressing male intellectual/contemplative hearers in the first instance. It is proper that any rhetoric of humility and dispossession should be subjected to suspicion. But this does not wholly turn aside the force of the Augustinian/Bernardine commendation. The style of talking about self-knowledge here discussed assumes that the most pervasive false construction of the self is an ego around whose specific satisfactions the world is to be structured. Hence: (i) this analysis is empty if it is not a tool for the questioning, by the disadvantaged as well as by the powerful themselves, of illusory (and thus oppressive) constructions of the world, (ii) if liberation is not to be a mere reversal of master–slave relationship, it must recognize what it is in human self-perception that generates and entrenches illusion and slavery, and (iii) liberation as overthrowing the bondage of the other's (oppressive) desire remains determined by that desire so long as it remains primarily negative and does not move on to address the question of how we might *now* imagine shared satisfactions. In other words, victimage is a dangerous rhetorical instrument, if it means that a language of primitive innocence violated is allowed to distract attention from the vulnerability of all historical schemes of self-presentation, particular and corporate, to the seductions of self-finalizing, closure to criticism. The recovery of an oppressed, victimized history is a profoundly necessary moment in so many enterprises of self-knowing—whether black history, workers' history in Britain, or the individual and terrible histories of abused children. Isolated from any energy in asking how present reciprocities might be turned towards shared goods, how desire becomes more than desire for the end of the oppressor's desire, the language of the victim can become sterile and collusive. The personal is not the political if it is stops at being a programme of negation and the reinstatement of an injured ego; and the political remains tribal if dominated by *ressentiment*. In establishing this, the project of a self-knowledge that emphasizes contingency and the non-finality of our 'constructions' of selfhood may have its place, even when all allowances have been made for the danger of this language in its turn.

Two concluding reflections. The first has to do with the more strictly philosophical import of all this—and, less directly, with the popular contemporary rhetoric of self-discovery. If we were to ask, 'How might we "test" for self-knowledge in ourselves or others?' it looks as if the answer might lie in trying to deal with questions like, 'Is there a pattern of behaviour here suggesting an unwillingness to learn or to be enlarged?' or 'Is there an obsessive quality to acts of self-presentation (in speech especially) that would indicate a fixed and defended image of needs that must be met for this self to sustain its position or power?' or 'Is there a refusal to deal verbally or imaginatively with the limits of power—ultimately with mortality?' In other words, we do not look first for acquaintance with any particular vocabulary of 'self-analysis' (we don't test for *information*). This may be a rather banal observation, so philosophically obvious as not to need saying; but in a culture where self-help books about self-knowledge, not least of a religious tinge, abound, we may well need reminding that a person may be possessed of a fluent vocabulary, well able to plot him- or herself on the charts of temperament and attrait and to retell their biography in the idiom of fashionable psychobabble, and yet continue to act in a way that seems to deny the recognition of mortality and the necessary ironies that go with it. And in so far as the present vogue for a religious rhetoric of self-awareness relies heavily on this kind of technology of redescription, it is indeed at odds with what the Christian spiritual tradition (and others) has meant by self-knowing. This is, of course, perfectly compatible with saying that, when we recognize a crisis of truthfulness, the power

225

of our habitual self-deceits, there is a place for theories of trauma, repression, the characteristic patterns of personality type or whatever, in unblocking certain channels and diagnosing the scale of our defensiveness; all this (a good servant and a bad master) has its role in becoming reacquainted with our contingency, even if it cannot deliver everything. The religious believer and the analytical or therapeutic theorist, however, will have different things to say about what more than theory is required.

And last *when* is the injunction, 'Know thyself' likely to be uttered, and *who* has the authority to utter it? When it can be shown that my actions are at odds with what might be expected of an agent both reasoning and mortal. When King Lear's daughters agree, 'He hath ever but slenderly known himself', they are pointing to the tension which his behaviour exhibits between verbal recognition of mortality and the obsessive clinging to the image of a royal self. Yet their own frightening egotisms disqualify them from having the right to execute judgment on his self-deceit. His eyes are opened in two ways: by the naked madman on the heath and by Cordelia; by Poor Tom to mortality and impotence, by Cordelia to the need of love. Who or what can command us to know ourselves? The dispossessed life—whether Tom's utter lack of standing and pride (itself in the play, of course, a *strategy* of dispossession on the part of Edgar), Cordelia's abnegation of revenge ('No cause, no cause!').[13] The injunction is there for us in the way in which the holy life interrupts our habitual constructions (an echo of Kierkegaard's *Philosophical Fragments* here)[14] by making for me in the world the room I thought I had to conquer and possess.

References

Ferry, L. and Renaut, A. 1990. *French Philosophy of the '60's. An essay on Antihumanism* (Amherst).

Girard, R. 1987. *Things Hidden Since the Foundation of the World* (London).

Lacan, J. 1968. *The Language of the Self. The Function of Language in Psychoanalysis*, trans. with notes and commentary by Anthony Wilden. (Baltimore).

[13] On Lear as a text about knowledge and self-knowledge, about the knowledge we need and the knowledge (of mastery, of information) that must be foregone in the process of moral maturation, even salvation, see the brilliant essay of Stanley Cavell, 'The Avoidance of Love: a Reading of *King Lear*', in *Must We Mean What We Say?* (Cambridge 1976), pp. 267–353. My debt to this and other works of Cavell will be evident.

[14] *Philosophical Fragments/Johannes Climacus,* ed. and trans. Howard V. Hong and Edna H. Hong (Princeton 1985), Part II, esp. pp. 30–5, and Part III, pp. 39–46.

Laing, R. D. 1960. *The Divided Self* (London).
Robinson, J. (ed.) 1977. *The Nag Hammadi Llbrary* (Leiden).
Vesey, G. 1991. 'Self-Acquaintance and the Meaning of "I"', *Inner and Outer. Essays on a Philosophical Myth* (London: Macmillan).
Williams, R. 1982. *Resurrection. Interpreting the Easter Gospel* (London).
Williams, R. 1992. 'The Paradoxes of Self-Knowledge in *De Trinitate X*', *Collectanea Augustiniana* (forthcoming).

Facing Truths:
Ethics and the Spiritual Life

MICHAEL McGHEE

I

In this paper I continue an enterprise begun in earlier work (McGhee, 1988, 1989) in which I attempt to naturalize into a western philosophical context concepts that derive from the practice of Buddhist meditation. In particular I shall try to make use of the notion of *samādhi* (sometimes translated as 'concentration') and *vipassanā* or insight. I should stress that I make no attempt at a scholarly explication of these terms but try rather to establish a use for them through reflection on experience, and by making a connection with concerns from aesthetics about expression and intentionality: I do so as a moral philosopher seeking to retrieve the Greek virtues of continence and temperance, which I have tried to relate to stages in the emergence of what I call an 'ethical sensibility', so that temperance, for instance, is the natural state of one in whom such a sensibility is flourishing. But I see the development of that sensibility as the *concentration* or gathering of a person's energies into its structure, into the sustaining of the thought or perception upon which action or non-action depends, as well as into the sustaining of action itself. In talking of 'energy' here I am trying to develop an idea of Simone Weil's in which she refers to 'the energy available for action'.

Not everyone is comfortable with the phrase 'spiritual life', perhaps for good reason, but I am using it for want of a better, and hope that I can draw attention to a set of traditional associations that will temper the discomfort. The point is to *track* its application rather than assume what it must be. In what follows I shall sketch a naturalistic conception of ethics in which I claim that moral dispositions are expressions of determinate stages of our spiritual life, including a stage, for instance, in which a person comes to feel 'a spirit which delights to do no evil'.

So I do not rely on a well-understood notion of the spiritual life in order to build up a picture of ethics, nor again on a well-understood conception of ethics to illumine the idea of the spiritual life. They are both thrown into contest and I try to construct the content of each out of the other. Even so I hardly approach what might be called the metaphysics of the spiritual life, the questions about what insights might be yielded about the nature of reality, even though I rely to some considerable extent on the notion of 'concentration' and the regulative idea of 'seeing things as they really are'. Despite the absence of metaphysics I do not want to reduce the spiritual life to ethics. On the other hand, in more than one tradition a connection is made between them, so that spiritual progress is manifested in a person's demeanour, or in the forms of relation established in a community. By the same token, there has been a traditional tension between the idea of spiritual progress and conventional morality.

To make a beginning, I shall present the spiritual life as a series of transformations of the persons of a community, from one set of dispositions and forms of connection (or disconnection), to another, in which there is no disconnection. It would be proper at this point to include the idea of a political transition from a social life governed by violence and domination towards one governed by freedom and justice, but though such issues provide a more or less unspoken background, I want to concentrate on moral psychology and its relation to ethical foundations, so that I can establish an *epistemological* approach that makes sense of the importance attached to such political values. The transformations I have in mind need to be shown to depend upon some notion of discernment or understanding, upon some recognition or acknowledgement of reality. Neither change nor *discernment* seem to me to depend directly on the will: in fact they seem to come to us as a kind of grace of nature. To capture the connections between change, understanding and reality I am inclined to talk of *transformations of sensibility,* if the term 'sensibility' may be thought not to recede too much from the idea of action and forms of contact. It is not so much that inner processes stand in need of outer criteria as that the single process by which the concentration of life and conduct occurs has inner and outer moments, and it is, finally, by our fruits that we shall know one another. But at least a rich enough conception of sensibility gathers together the ideas of responsiveness and knowledge, motivation and cognition, the idea of 'reading' and acting in its light, as well as the idea of false readings, misrepresentation, delusion and ignorance. So I shall retain the notion of sensibility, despite the suspicion of 'feeling' to be found among philosophers, because it carries with it the idea of being *moved* by what one has come to know, see, or believe, an idea without which action, even with knowledge, is inexplicable, and because feeling, as Sartre recognized, is revealed in action. The key idea is that of the transforming power of knowledge or understanding, itself dependent upon a sufficient degree of alertness or concentration. Touched by reality, we are affected, and act.

I shall borrow a phrase of Schiller's to express the epistemological *leitmotif*: 'impressions move the soul'. The status of claims about what

kinds of impression move to what kinds of response is really the central
topic. I think there are truths to be established about such matters, and
that our knowledge or ignorance of how impressions move the soul
provides the grounds of mutual expectation, the grounds of our sense of
how a person ought to behave, of what attitudes they *ought* or *ought not*
to have. But when I talk here of what attitudes a person ought or ought
not to have, I do so not in the sense of some fugitive moral 'ought' or
unexplained form of normativity or obligation, but in the sense of *what
we have reason to believe or expect.* The epistemic expectation that I
have in mind sits well with the expressivist position that underlies my
claim that our ethical dispositions are natural expressions of determinate states of consciousness, or stages of 'concentration', which I understand as the suspension within which a form of perception and conduct
are capable of being sustained. Our mutual expectations are grounded,
in other words, in implicit beliefs about the intentionality of sensibility
or motivation, about *what* impressions move the soul, and in *what* way.
Any community's expectations will be limited by the range of responses
that are so far available to it, by its known sensibility, possibilities
which can be lost and found in the tradition, again and again. Known
motivation provides the ground of expectation in a way that matters to
us, sometimes urgently, hence the illusion of normativity associated
with expectation, the anger, for instance, at its disappointment, when
in reality the expectation is merely empirical. Relative differences in
our knowledge or ignorance of how impressions move the soul is also
the source of mutual incomprehension, though the incomprehension is
not always mutual. As Kierkegaard observed, two people can agree
with one another, word for word, and yet be guilty of the grossest
possible kind of misunderstanding. We cannot neglect this notion of
incomprehension, since it seems to be an unavoidable feature of the
notion of *progress* or, more neutrally, states of concentration, which
define an available horizon, bring into focus particular forms of motivation, sustaining the associated thoughts and perception which move to
action. Implicit in this latter thought is a theory *about* ethics, that
conduct, and forms of community, depend upon degrees of concentration of energy.

I have been reminded by recent work of Nicholas Lash (1987) of the
pull of a tendency to which I have already to some extent succumbed by
my readiness to talk of impressions moving the soul. It would be natural
to proceed at this point somewhat as follows. The impressions that we
perceive are perceived under descriptions, and it is in the light of our
discernment of particular realities and their causal connections, represented thus and so, that we are moved to action or restraint, though to
what action, and by what impressions, depends on the state of the

subject, an aspect usually neglected in discussions of emotion. To say this is to make a quite proper gesture towards the intentionality thesis as it operates in relation to the emotions or sensibility, and it marks the cognitive aspect of feeling: we respond to realities as we take them to be. But though this may suffice as a very general indication of that aspect of sensibility, we need some strong qualification if we are to characterize the nature of our ethical responses. *The tendency, of course, is to swerve away from the personal*. If we say that impressions move the soul, it is easy to refer that to facts, to states of affairs, to situations, to the objects of the emotions. There is nothing wrong with this, in fact it is something to emphasize, but there is an essential rubric under which such talk should fall. We are not just affected by what happens in the world, though we are: we are affected, under certain conditions, which I intend to discuss, by how persons, and, more generally, fellow-creatures, are affected by what happens in the world. Such natural patterns of response conflict with egocentric or communal patterns of thinking in which events are related exclusively to oneself or one's own community. But even in our own case we respond to states of affairs *as they affect us*, (and they affect us according to how we are then constituted), either in immediate reaction, or through some conception of ourselves as persons among others, which may or may not be adequate to how we are then constituted. Thus *we act for our own sake*, perhaps to remedy a situation by which we are adversely affected, or see ourselves as adversely affected.

The ethical life, though, opens up at the point where we respond to situations as they affect persons *qua* persons, and then, more generally, as they affect other sentient beings, and where, in consequence, *we act for their sake*, in accordance with our understanding of the nature of their being. So impressions move the soul, but it is our sense of what is happening as it affects one another for good or ill that moves us to *ethical* action. And we cannot avoid the issue that our sense of what is harmful or beneficial is dependent on what we take ourselves to be, upon our sense of the nature of our own reality. To put it another way, with an obvious Kantian resonance, in the ethical life we become for each other *ends of action*. What I mean by this is that we come to want to sustain each other in our own being, a being which I take to be essentially ethical. But our conception of what we are may or may not coincide with the *realities* of personhood. For instance, our view of what it is to be a person may fail to acknowledge the truth of the claim, if it is a truth, that unless a grain of wheat fall into the earth and die, it shall not bear fruit. I say 'if it is a truth' because I take it that individuals stand in different relations to this remark, and others like it, from the point of view of their experience. That they do so makes the issue of moral realism difficult in unexpected ways.

The background motivation is not an intellectual or so-called 'ethical' commitment to the flourishing of persons. On the contrary, in the spirit of my attempt to offer a naturalistic reading, I should say that the background motivation emerges from the *finding*, and the finding satisfaction in, possibilities of mutual response, solidarity, and other notions constructing the arrival of an 'existence for one another'. All these notions could be summarized under that of the *appreciation of persons*. But the arrival of these possibilities *determines the self that comes to be constructed out of them, and which did not previously exist*. To put it another way, the terms in which we become for one another 'ends of action' depends crucially upon the realities associated with 'appreciation'. Under certain conditions, of *meeting* and *union*, which bring about this 'existence for one another', we come to appreciate the *beauty of persons*, and are thereby regenerated in a way which should, rationally, be reflected in our conception of personhood. This appreciation provides a background sense of what persons are that can become the ground not only of delight but also of compassion. It can become such a ground for an important reason. Our *sense* of what is harmful or beneficial, of what we can rejoice in or feel compassion for, is dependent on what we take ourselves to be. But if we are constituted by the trajectory I am attempting to describe, then what diminishes its possibility we shall treat as 'harm' and what enhances it as 'benefit'. I refer to a trajectory, and that allows us to say that there is a time before and a time after we come to appreciate the beauty of persons and to treat them as ends, a time before and a time after, that is to say, the emergence of *ethics*. And since, as Diotima insists, we desire to procreate in beauty, it is out of the reciprocal appreciation of the beauty of personhood, and hence of its possibility within one another, that forms of life are constructed in which persons are let be and enabled to become.

Since I am making central to my account this Platonic notion of appreciation of beauty of soul and the associated desire to procreate in beauty, I had better rehearse how I think such beauty can be the object of desire, or rather, how certain states of persons can be considered 'beautiful'. Someone who *exemplifies* such beauty embodies or mediates a certain concentration of energy, that by which a certain perspective and demeanour is sustained, a perspective and demeanour that become the *form* of that mode of energy. By 'perspective' here I simply mean the being moved in a determinate direction by particular kinds of reason, i.e., states of affairs that are the intentional objects of particular motivations. Coming to treat persons as ends is a matter of coming to be established in the corresponding ethical sensibility focused on well-being. As I have just said, the sustaining of such a perspective, such a sensibility, depends upon a concentration of energy around the relevant forms of attention. But such energy *radiates* and attracts, is an object of

Eros, a form of beauty. I have suggested elsewhere (McGhee, 1990) that one reason such forms attract, are the object of *Eros*, is that they represent the direction of our own *Bildung*, our own future. I had better add, since there are false teachers and false prophets, that the energy radiates in demeanour and disposition, which provide the criteria of identity for forms of energy. However, the *conjunction* of such energies, between persons, produces new forms, a new spirit, hence my reference just now to the experience of meeting and union, and these *discoveries of contact* produce transformations that rationally determine changes in our conception of personhood and hence of what we take to be harmful or beneficial.

Many philosophers, including Thomas Nagel and Simone Weil, have tried to make the notion of 'recognizing the reality of persons' work for them in this context. But such phrases are not really helpful unless we tie them to something more fundamental which determines the *form* of that recognition. What I suggest is that this recognition is manifested in the discovery of the forms of relation and construction that constitute the ethical life and determine our sense of what that reality is. Simone Weil writes that 'belief in the existence of other human beings is *love*'. But I would rather say that love is a *form* of belief in the existence of other human beings, one moreover which helps to construct the nature of that existence.

It is crucial for my later argument that it should be a common experience that when someone finds this perspective and form of contact and loses it again they experience the loss as a darkening and oppression of the spirit, in a way analogous perhaps to the expense of spirit that occurs in sexual excess. We should associate this darkening of the spirit, not merely with the sense of energies scattered and concentration and focus lost, but with its corollary, what I would call a loss of *world*, a closing in of the horizons, depletion, isolation, envy, hatred, which again is imaged in the sexual life by the experience of lust as 'a drug against imagination of all but carnal forms' (Morgan, 1936, p. 267); and we may contrast this with the sense of energies returned and gathering, increase in intensity answered in extensity of scope, enlargement of sympathy, responsive to a touch, a glance. The ebb and flow of spirit is the epistemological ground of rational expectation in regard to conduct. The ebb and flow is between what Blake calls the two contrary states of the human soul, between what Rousseau calls the state of nature and the just society. The claim is that recognizably moral dispositions reflect some state of the soul, or better, some state of a society. But I want to offer an epistemological account of what I have just called moral dispositions, a moral epistemology. The experience of the ebb and flow between hatred and love, that is to say, the common experience, attested to in our traditions, provides the rational grounds of expectation, in the sense that even in the palpable absence of ethical response there are grounds for the judgment that the response ought to be there. But this ought is to be given an epistemic construction in terms of rational expectation.

Kant talked of treating humanity as an end, whether in one's own person or in that of any other. I think he was mistaken to think in terms of an imperative in this context, as I shall explain later. But I want to retain the reference to 'ends', without subscribing to the conception of 'humanity' under which it was supposed to happen, and I want to relate it to a characteristic sensibility and form of life, as I said. These come about under certain conditions, and their emergence or growth occurs against the radio interference of forms of life that are already in place: hence the two contrary states of the human soul. But for Kant the failure to treat humanity as an end is represented as a matter of treating it merely as a means. No doubt there is such a contrast, but it seems to me that a fuller account of what we could call the 'anti-ethical' must include the active hatred of humanity, the contrary of appreciation, and hence an attitude specifically directed towards beauty of soul, the attitude of Claggart towards Billy Budd, for instance. Nevertheless if we think in terms of the two contrary states of the human soul we can represent the polar possibilities as the active or passive hatred of humanity, on the one hand, and the treatment of humanity as an end, on the other. But as I have already said, it is a particular development of *Eros*, our coming to appreciate the beauty of persons, which determines the formation of what I am calling the ethical life, and, consequently, our coming to treat one another as ends of action, as beings we desire to sustain in a being we mutually create. The discovery of this general form of relation marks the transition between *the* two contrary states of the human soul. Blake does not refer to two among a number. These poles represent the *natural possibilities*, a claim important for any moral epistemology, and they correspond to polar forms of *sensibility*, the passage from the one to the other of which is arduous, a matter of struggle and hazard, away from the reactions of violence, towards the feeling of a spirit that delights to do no evil. Responding to one another as ends, in the terms I have described, refers directly to a distinctive sensibility and form of life, *must* do so, if it is not to be depraved into the mentality of the *Collector*. For instance, coming to appreciate each other's beauty as persons, we are *solicitous* for one another about the good or harm that can befall us. But Diotima connects that reciprocral appreciation to procreating in beauty, as we have already seen. What are begotten or constructed out of such meetings are beneficent forms, the co-operative building of forms that aid each other's free development.

Now once we recognize the passage from one pole of sensibility and structure to another, we need to put content into the movement from the anti-ethical to the ethical by putting in place a multiplicity of concepts which structure the realities of this movement. In other words, if we are to talk plausibly about the realities of personhood, then we need to show their structure. I have in mind, for instance, such concepts as consent, the conditions under which it is given, trust, reconciliation, forgiveness, and so forth. The point I want to make is that these represent *realities* whose nature we have to learn, whose nature we may not immediately grasp, because they have to be learnt from experience. The processes of forgiveness, for instance, have their own reality independently of what any individual may happen to think, so that here there are truths we know or fail to know. Whether we are interior to these realities or not determines the extent of mutual understanding, but, more importantly, our knowledge or ignorance determines our sense of what enhances or diminishes our personhood, since it determines our sense of what *constitutes* our personhood, so that the direction of our solicitousness for someone's welfare may *collide* with realities that we do not understand, in such a way that the well-meaning become enemies of the good.

II

What I have presented so far is the barest sketch of a transition towards the ethical life, which I have represented in terms of discovery, construction and education, one in which we are concerned for each other's welfare, qua persons, under a particular conception of personhood, one that needs rationally to keep pace with the transformations of sensibility by which we are constantly reconstructed. The forms of that relation to one another, their realities, already provides us with some sense of what constitutes our personhood, since they serve, or help, to construct it, and our knowledge or ignorance of the relevant truths determines our understanding of what enhances our being or diminishes it, though attachment to particular constructions may blind us to the further reality that in order to find our life we must be prepared to lose it.

However, in this section, I want to enter into a more detailed discussion of what constitutes moral truth, or how we are to understand the way questions of truth enter into ethical discourse. I have already promised that I would carry out an epistemic reinterpretation of 'ought', and I turn to that now. What I offer is intended as a general account that is neutral between different conceptions of the good or rival moralities.

One of the issues in contemporary moral philosophy is whether moral utterances that have the form of statements have a substantive factual content. If they are genuine statements then they should have a truth value. But before we can decide whether or not they do, we need to know what are to count as moral statements in the first place. Thus, for instance, the presence of key terms like 'good', 'bad', 'ought' and 'should', 'right' and 'wrong', too readily identified as 'ethical symbols', do not by themselves establish that a purported statement is a moral one. (The current use of 'wicked' as a term of commendation should make this clear). For example, the claim that you shouldn't beat your children does not yet give us the information we need to identify it as a moral claim. This becomes obvious when we turn to the speaker's *grounds*. If the coupling of the claim with its grounds yields the statement that you shouldn't beat your children because you are likely to get prosecuted, then it is clear that we have a prudential judgment.

Let us stay with this prudential judgment for a moment and draw out the way it might function as a hypothetical imperative. In addition to the reason for the judgment—you are likely to get prosecuted—there is an implicit parenthesis that says, 'and you don't want that, do you?', that is to say, an assumption about the wants or desires of the hearer.

But there are two ways in which the judgment is defeasible. It may be false that you are likely to get prosecuted, so on *that* account it is not true that you should stop beating them, though there may be other facts which would render it true. There is, of course, a second hurdle. If the assumptions about the hearer's wants are wrong, if they *want* to get caught or are indifferent to that outcome, then again, on *that* account, it is false that they should stop beating their children, though again there may be other desires that make it true.

But now, if the reason offered for the judgment that you ought not to beat your children is something like, 'it will traumatize them' or 'do them harm', then this grounding reference to their welfare points towards the utterance's being a moral statement. The moral statement is not, 'you should not beat your children', but rather, 'you should not beat your children because it will do them harm'.

The familiar crux, though, is whether we need to add the same implicit parenthesis that we applied in the prudential case. Do we need to add to the reason that it will do them harm the assumption that the hearer has some relevant moral end, without which the judgment has to be withdrawn because a necessary condition of its application is missing? The question that is familiar to moral philosophers since Philippa Foot's influential article is whether so-called 'moral oughts' are similarly hypothetical upon the ends of the agent. Does the *truth* of the claim that you should not beat your children depend not only on the truth or otherwise of the grounding judgment that it will harm them ('it never did me any harm'), but also on the truth of the hypothesis that you don't want them harmed? In other words, does the fact that you

want to harm them, or are indifferent to their being harmed, undermine the truth of the claim that you should stop beating them? Does the
judgment apply to you only if you have the relevant moral end of
causing none harm?

If you want to harm them, or are indifferent to their being harmed,
then it looks as though we should say that it is *false*, at least on that
account, that you should not beat them. On the other hand, I shall
introduce an epistemic construction of 'ought' at just this point, indicat-
ing an *expectation* that derives from beliefs about what impressions
move the soul: the causing of harm being one such. What I want to
consider is whether it might be *true* that you should stop beating your
children if it is true that by beating them you cause them harm *and* if it
is true that you *ought* to want not to harm them. The thought here is
that the reference to their well-being or harm draws attention to the
intentional object of a form of motivation that the speaker expects to be
present in the hearer. If the thought is correct, then the practical ought
reduces to a *deployment* of the epistemic ought. It is in the expectation
that they are thus motivated that the attention of the hearer is drawn to
the relevant fact in the first place. What I am suggesting is that in the
case where the hearer *denies* that they are thus motivated and the
speaker declares that they ought to be, the speaker is claiming that there
was reason to believe that they would be on the grounds that people are.

Now clearly there are going to be difficulties with this claim. In
particular it is important that it should not be the means of smuggling
into the account an unacknowledged normativity. In a moment I shall
make some important qualifications. Before coming to that, it is worth
pointing out that someone who makes such a remark will be aware that
the claimed indifference represents a state of affairs that may be dangerous from the point of view of well-being, and drawing attention to a
disappointed expectation is a possible means of eliciting or awakening
the expected motivation.

So what is the force of the claim that you ought not to want to harm
people? The first thing to say is that *'ought' does not here govern an
action but a want or desire*. It makes reference to the idea of a desire
that someone ought to have, in the sense, I shall maintain, that there is
reason to believe that people do have that desire, or would have it under
certain conditions.

Clearly there is an important difference between the belief that
people do have a certain desire, and the belief that they would have it
under certain conditions. In order to clarify the difference, and explain
its significance for the position I am trying to develop, I shall invoke a
context of innocence and a context of experience.

The naive version of the epistemic construction treats the moral
derelict as anomalous or weird. In fact this is the form of a very common

reaction to disappointed expectation in moral contexts. After appealing to certain considerations that the derelict claims to be indifferent to, many people may think or say, 'but you *shouldn't* be indifferent', and I suggest that their meaning is something like, 'but people *aren't* indifferent to such things: you're weird'. I think it is plausible to claim that this is the structure of a common response in such circumstances, it is an expression of incredulity in the face of recalcitrant experience. The (sheltered) background belief is that people as a matter of fact respond in particular, predictable ways, and here is someone who doesn't, someone anomalous or weird. The utterance may, of course, carry all the weight of fear and loathing the context may elicit. Disappointed expectations cause dismay because they make a difference.

But though I think it is plausible to claim that many people do in fact respond in this way, it is clear that the response itself is inadequate. The background belief is manifestly false. The more brutal the environment the innocent find themselves in the less likely are they to retain that belief. On the contrary, it is all too likely that brutality becomes precisely what they expect, and they may come to look askance at its absence rather than at its presence.

Let me now with some caution invoke a context of 'experience'. I have already alluded to the idea in an earlier part of this paper. My proposal is that experience perhaps may lead someone to a more sophisticated background belief, viz., people are moved by 'ethical' considerations, *unless they are prevented*, if I may so formulate it, at least provisionally. So when such a person addresses the derelict, saying 'you *should* be moved', knowing full well that they won't be, we should understand them to be saying, 'you *would* be moved by such things if you were not prevented'. In other words, they *expect* the relevant attitude to be missing, and they have an explanation.

I said earlier that I thought Kant was mistaken to think in terms of a categorical *imperative*, and my thinking so is connected with the possibility of treating 'ought' epistemically. Kant writes plausibly enough about the need of imperfectly rational beings to address imperatives to themselves, but it seems to me that he becomes confused when he associates prescription with 'ought'. He claims that 'all imperatives are expressed by an "ought"' and this because they are addressed to a will which is determined by other considerations than the good, and so has to be necessitated (hence the imperative). In the case of the holy will on the other hand,

> there are no imperatives: '*I ought*' is here out of place, because '*I will*' is already itself necessarily in harmony with the law. Imperatives are in consequence only formulae for expressing the relation of objective

laws of willing to the subjective imperfections of this or that rational being—for example, of the human will. (Paton, p. 78)

But there is only a contingent relation between the presence of 'ought' and the need for self-addressed imperatives. 'Ought' as a logical indicator of a practical judgement applies as much to the holy will as to the imperfectly rational will, and simply indicates the presence of a reason for acting. Once a reason has been established or become apparent, the holy will is no doubt determined appropriately, says 'I will' and acts. This is not the case with the rest of us, but this fact simply reveals the imperfect human context in which ought-judgments are uttered, as it were, through clenched teeth. In any case it seems plausible to offer an epistemic reading of the so-called imperative. To claim that as rational beings we ought to act in such a way that we always treat humanity, in our own person, or in the person of any other, never simply as a means, but always at the same time as an end, is in effect to claim that there is evidence that this is what rational beings *would* do, if they were not prevented by the sway of contrary inclinations. We *are* moved to treat each other as ends—unless we are in some way prevented—and to that extent, of course, our actions are in conformity with universal law, in conformity, that is to say, with a conception of what any human being is.

I shall turn aside from this discussion of the epistemic ought for a while because it needs to be connected with another issue, that of the meaning of terms like 'right' and 'wrong', which I shall treat as expressive, in certain contexts, of particular forms of motivation in relation to their intentional objects.

But in order to consider how to construe these terms I shall discuss another aspect of moral judgment that is also connected with the issue of truth and falsity.

What may tempt a philosopher to think that moral statements possess some sort of 'mind-independence' is the fact that there is sometimes a determinate answer to the question, about some proposed action, 'what *makes* it wrong?' Furthermore, a child seeking reassurance, who exclaims, 'it's *true* that it's wrong, isn't it?' may reasonably be given the same determinate answer.

But this reference to some action judged to be wrong is hopelessly under-determined. So let us look at an example. Simon Blackburn offers an apparently promising one: it is wrong to kick the dog, and he says,

it is not because of the way we form sentiments that kicking dogs is wrong. It would be wrong whatever we thought about it. Fluctuations in our sentiments only make us better or worse able to appreciate how wrong it is. (Blackburn, 1984, p. 217)

So far as fluctuations in our sentiments go, the intentionality thesis allows space for the *rooting* of them in specific states of affairs, their objective correlative, which is also a criterion of their identity. If our emotions are unstable, the criteria of identity are not, for they represent the cognitive element which a person may hang on to despite the ebb and flow of affect. Though the thought that such and such an action would be *wrong* may be fairly drained of affect, it is nevertheless the channel along which feeling flows, and even the heroic, willed refusal to commit the act draws strength from the motivating description, and represents a modification of feeling, feeling as it is under certain conditions. The idea that our sentiments are subject to *fluctuation* reflects, of course, a particular, contingent experience of them. However unstable what we call our feelings are, their counterparts in the language, the evaluative *foregrounding* or *isolating* of specific phenomena, by which we give them expression, is itself relatively stable. The relationship between evaluation and description reflects the intentionality of motivation.

But Blackburn's comment seemed to provide some grounds for a 'quasi-realist' position on moral statements: it seems to be true that the action is wrong whatever we think—or feel—about it. It also helpfully provided a *caveat* against a too simple view of the implications of asserting a 'subjective source' for ethics. He goes on to give the obvious answer to the question, 'what makes it wrong?' and it is the answer one might well give to the child who wants to be reassured that it is *true* that it is wrong to kick the dog: 'what makes it wrong to kick the dog is the cruelty or pain to the animal'. If we wanted to reassure the child we may well say that it is true that it is wrong to kick the dog because doing so causes pain to the animal.

But is this really such a satisfactory example of a moral statement whose candidature for truth or falsity we can now go on to discuss? The example seems to imply that there is something in virtue of which an action is wrong, something, moreover, that makes it wrong whatever any particular individual may think or feel. In fact it is unilluminating. It does no more than to show the relation between one description of an action and another, one that presents it precisely as the focus, the intentional object, of a particular sentiment. Kicking dogs is wrong because it causes them pain, and the causing an animal pain is something there is some reason to expect that anyone will recoil from, though as we know very well the expectation may be disappointed.

Presumably, though, we should want to say, not only that it is wrong to kick the dog, but that it is wrong to cause animals pain, or needless pain, which is, after all, what is wrong with kicking the dog. But if it is wrong to cause unnecessary suffering, or, to use an example of Timothy Sprigge's, wrong to cause pain to others for one's own enjoyment, then *what makes it wrong*? I present that as a rhetorical question, since I do not think the case is parallel with asking what makes it wrong to kick the dog. That question, I suggest, demands to know what the relevantly motivating description is, a motivation whose general direction is *expressed* in the use of 'wrong' in connection with causing unnecessary pain.

But if what I presented was merely a rhetorical question, can we not ask, nevertheless, whether it is *true* that causing unnecessary suffering is wrong? Some philosophers would simply treat it as an emotive utterance, not to be taken as a genuine statement at all. Others may see it as a truth to be discerned by rational insight. Others again may see it as the expression of a principle we simply have to choose or commit ourselves to.

My own position is closest to emotivism, insofar as I agree that the statement draws a kind of exclamatory attention to the intentional object of an emotion. But it is to an emotion, as it were, that we expect to find. And here we must return to the discussion of the epistemic ought, so that I can offer at the same time a naturalistic reduction of so-called ethical statements *and* an emotivist reading of such terms as 'wrong'.

What qualifies my genuflection to emotivism is that while I think that 'wrong' gives expression to a response, it is to a response that *anyone ought to have*. What I mean by this is, a response that there is reason to believe that anyone *will* have. Let us leave on one side for the moment the necessary qualification 'unless they are prevented'. We are initiated into these expectations, learning profiles of personhood in the acquisition of language.

It might help to make this plausible if I point out that when a speaker says something like 'you shouldn't kick the dog' and offers the reason, 'because it will cause it pain', they do so to remind, or bring to the attention of the hearer, a *fact* which they expect to move them, a fact, that is to say, that they think *will* move them, and this thought is grounded in the background belief that people *are* moved by such considerations, a belief which gives point to the attempt to draw attention to the relevant fact, an attempt premised on the assumption that anyone will be moved, or, with experience, may be moved, or may just possibly be moved. To exclaim that it is *wrong* expresses, as do tones of voice, etc., the nature of that motivation, but it is essentially a motivation that anyone ought to have in the sense I have indicated.

I think the implication of this is that to claim that some form of conduct is wrong, in the context under discussion, implies that it is a form of conduct that anyone ought to avoid, and this is to be construed as: a form of conduct there is reason to believe anyone *will* be moved to avoid. The advantage of this analysis, incidentally, is that it allows for non-expressive uses of 'wrong', including its use in non-asserted contexts. Thus when I say x is wrong I may or may not express the attitude I thereby imply that anyone ought to have. In unasserted contexts, where I say, 'if x is wrong, then . . .' I am saying 'if x is such that anyone ought to avoid it, then . . .', which is to say, 'if x is such that there is reason to believe that anyone will avoid it, then . . .'.

If what I say is more or less right then we need to part company with the simple emotivist analysis of 'it is wrong to cause unnecessary pain'. What I offer instead is 'causing unnecessary pain is something anyone ought to be moved to avoid', which is to say, 'causing unnecessary pain is something there is reason to believe anyone will be moved to avoid, unless they are prevented'.

Part of the interest of a quasi-realist view of moral statements was to yield statements which, though subjectively derived, nevertheless made a claim to truth independent of what any individual happened to think. In the previous section of this paper I have already drawn attention to possible truths which are stronger candidates for realism. Thus I claimed, for instance, that there is a truth about the conditions under which forgiveness is possible, which holds firm quite independently of what anyone happens to think of the matter, as is the case, say, with 'fire burns'. If we ignore either we are likely to get our fingers burnt. The sort of truth I have in mind enters into one's conception of good and harm, and is significant for the development of a rather different moral realism from that to which we have become accustomed. But in the present context we are discussing another candidate for truth, the claim, e.g., that people are moved to avoid doing what causes harm, unless they are prevented.

In the first part I said that I wanted to sketch a conception of ethics which showed our moral dispositions to be expressions of determinate states of our spiritual life. The proposition that there is a determinate state in which we would naturally recoil from cruelty, for instance, is certainly truth-claiming, and obviously underpins the evaluations which give expression to it. To the extent that it is an expression of our being, we are harmed if the emergence of that state is suppressed. Such states of the spiritual life are a matter of the *sustaining* or holding together of the relevant sensibility, the holding in focus of the objects of awareness upon which action depends, and the maintaining of the possibility of action itself. The transformation of personhood is thus a matter of the concentration of energy within the relevant forms, forms that construct the person out of an ethical sensibility. The ethical sensibility becomes the form of the person, and the travails of its emergence are the travails of personhood.

However, the immediate claim is that anyone *would* recoil from cruelty, say, unless they are prevented. And the immediate problem is, how can it be defended? How might one begin to establish, or, more modestly, find evidence for, the proposition that someone who was indifferent to humane considerations was in some way prevented from being otherwise, someone who presumably wouldn't think of *themselves* as prevented?

I have given the gist of how one might go about this in my earlier remarks about the darkening and oppression of spirit which I said characterized the periodic loss of an ethical sensibility. What is crucial is the experience of an *oscillation* between two poles, between the two contrary states of the human soul, between the conditions upon which depend Cruelty, Jealousy, Secrecy, Terror, on the one hand, and Mercy, Pity, Peace and Love, as Blake represented it, on the other.

Whether there is such a trajectory from one state to the other is something that has to be personally appropriated from experience, the culture and the tradition. What has to be a matter of personal appropriation, a truth that can only be established subjectively, as Kierkegaard would put it, is the experience of an oscillation away from and in the direction of *a determinate orientation*, a tendency away from one form of life and towards another, so that our estimation of the poles is already stacked in favour of the *terminus ad quem*.

It is significant that the negative phases of this movement may be felt as a diminishing of the power of action, and a loss of world, because it entails the loss of the concentration of energy upon which such action depends, and the dismemberment of a formation of the self. Indeed one of the extreme points of oscillation involves the total eclipse of the perspective by which the contrary pole is constituted. Perhaps the eclipse is usually only partial, manifested in the experience of *incontinence*, which is an important clue. It represents an inability to act in the light of ethical considerations, even though one is moved in that direction. Incontinence represents a kind of partial eclipse, a moment between total eclipse and unimpeded vision. It is a state of mind which *prevents* someone from *acting* in the light of acknowledged good. I suggest that there are analogous states of mind which can prevent a person, not only from acting but also from being moved by the relevant impressions in the first place.

So the claim is that the derelict *would* respond if their vision were not obscured. It is the obscuration of vision that prevents the possibility of ethical action, that is to say, the *finding* of those forms of contact that determine our coming to treat persons as ends. I make that qualification because the image of obscured vision could be misleading: it might suggest that a sort of ethical vision is there all the time, if only the clouds were not in the way. In reality the obscurations prevent the *formation* of that vision.

So the claim that the moral derelict *would* respond ethically if they were not prevented amounts to saying that a certain formation of the person *would* develop if they were not entrenched in the states of mind that prevent it, not an arbitrarily privileged formation, *but the one we have found*. They are entrenched, we should have to say, in the known states of mind that prevent the insisting tendency that belongs to them as persons, a development by which they are constituted.

It is appropriate, I think, to draw attention to the precariousness of this claim. I have already said that whether we are constituted by this trajectory is a matter to be subjectively established, a truth to be ascertained between persons, and within oneself. It is hardly *well* established, there are too many instances where, if it is not discon-firmed, it is certainly not confirmed. On the other hand, people do change, and the presence of goodness has had observed effects.

So we rely on a view, to be subjectively established, about how we are constituted, a view about how anyone is constituted, to the effect that we are progressively structured around the trajectory I have described, around the emergence of an ethical awareness, and around the travails of its emergence.

If someone's *conception* of what a person is is derived from the experience of that transformation, a transformation that represents their *Bildung*, then their judgment that the delinquent *ought* to be moved by ethical considerations can reasonably be thought of as epis-temic, even if it is compounded by consequential evaluations. The delinquent ought to be thus and so because that is how people are, and how people are is a matter of the embodiment of the natural history of an ethical sensibility. In addressing the derelict in these terms, you should not be like that, the speaker attempts to recall them to themselves.

The corollary is that our sense of what is harmful or beneficial is dependent on what we take ourselves to be. If we are constituted by the trajectory towards the ethical life that I have described, then what prevents its possibility is felt as harm, and what furthers it is felt as benefit.

We thus have a rational conception of the good that undercuts much moral relativism while being compatible with wide cultural divergences in the expression of determinate spiritual states. This is a point worth highlighting because it is a common enough problem in moral philos-ophy that while morality is focused on the avoidance of harm or the furtherance of well-being, we have radical disagreements about what is to count as good, or count as harm. While there are wide cultural differences in the matter of what will bring about harm, or what will *be* harm, it seems to me that a rational limit is set on such conceptions by the fact, if it is a fact, that we are formed out of a sensibility associated with treating persons as ends, seeking to sustain them in their being. But such a state of *Eros* is their being, and if it is undermined they are harmed. *Treating persons as ends becomes a matter of sustaining them as beings who have precisely that aim, and the sensibility it entails.*

I am tempted to talk here in terms of our real constitution, but I think it is a temptation to be avoided. The implication would be that we are thereby fully formed and that there is some independent standard by which we could confirm as much. But maybe we are still forming. The experience of a progressive concentration around the formation of an ethical sensibility must already to some extent undermine the idea of a fixed self we need to hold on to, and perhaps a doctrine of *anattā* could be developed from the idea that the flourishing of that sensibility is dependent upon conditions. Among those conditions is the steady retrieval of energy from forms of conduct and patterns of reaction and thought that are already *unsatisfactory* (or *dukkha*) from the point of view of the emergent sensibility which already announces itself in pre-reflective unease and *post-mortem* remorse. What I have attempted to describe is a particular form of concentration, one which sustains the possibility of active thinking and the dawning of understanding. What a more sustained concentration may reveal about the nature of reality, and our relation to it, remains, perhaps, to be seen, for I speak, alas, as a philosopher, and not the exponent of a tradition.

References

Blackburn, S. 1984. *Spreading the Word* (Oxford).

Lash, N. 1988. *Easter in Ordinary* (London: SCM Press).

McGhee, M. J. 1988. 'In Praise of Mindfulness', *Religious Studies* 24.

McGhee, M. J. 1989. 'Temperance', *Philosophical Investigations* 12.

McGhee, M. J. 1990. 'Notes on a Great Erotic', *Philosophical Investigations* 13.

Morgan, C. 1936. *Sparkenbroke* (Macmillan).

Paton, H. J. 1978. *The Moral Law* (Hutchinson).

Notes on Contributors

Stephen R. L. Clark is Professor of Philosophy at Liverpool University. His most recent publication is *God's World and the Great Awakening*. Other writings include *The Mysteries of Religion*, *Civil Peace and Sacred Order* and *A Parliament of Souls*.

Sarah Coakley, who is Fellow and Tutor in Theology at Oriel College, Oxford, was previously Lecturer and Senior Lecturer in Religious Studies at Lancaster University (1976–91). She studied Theology at Cambridge and Harvard and is the author of *Christ without Absolutes: A Study of the Christology of Ernst Troeltsch*.

John Haldane is Reader in Moral Philosophy at the University of St Andrews. He is widely published in collections of essays and such journals as *Analysis*, *British Journal of Aesthetics*, *Inquiry*, *Philosophical Papers*, *Philosophical Review*, *Philosophy*, *Philosophy and Phenomenological Research*, *Proceedings of the Aristotelian Society* and *Ratio*.

Ronald W. Hepburn has been Professor of Moral Philosophy at the University of Edinburgh since 1975, and previously taught philosophy at the University of Aberdeen and the University of Nottingham. He was Visiting Professor at New York University, 1959–60, was the Stanton Lecturer in Philosophy of Religion at Cambridge, 1965–8, and gave the Margaret Harris Lectures on Religion at the University of Dundee in 1974. He has written *Christianity and Paradox* (1958, 1968) and *'Wonder' and Other Essays* (1984); he has broadcast and contributed articles to books and journals on philosophy of religion, aesthetics, philosophy of education and moral philosophy.

Fergus Kerr, who is a friar of the Order of Preachers, teaches philosophy at the Roman Catholic Seminary in Edinburgh. He is the author of *Theology after Wittgenstein*.

Oliver Leaman is Reader in Philosophy at Liverpool Polytechnic, and is the author of *An Introduction to Mediaeval Islamic Philosophy* (1985), *Averroes and his Philosophy* (1988) and *Moses Maimonides* (1990).

James Mackey is Thomas Chalmers Professor of Theology in the University of Edinburgh; he previously taught philosophy at Queen's University, Belfast. His most recent book is *Modern Theology: A Sense of Direction*, and he is currently writing a book on 'Power and Christian Ethics' for Cambridge University Press.

Michael McGhee, who lectures in Philosophy at the University of Liverpool, was formerly Lecturer in Philosophy at the University of East Anglia (1973–87), he has contributed articles to *Religious Studies*, *Philosophical Investigations* and the *British Journal of Aesthetics*; he is a member of the Western Buddhist Order.

Notes on Contributors

Janet Martin Soskice is a Lecturer in the Faculty of Divinity of Cambridge University. She is the author of *Metaphor and Religious Language* (1985).

T. L. S. Sprigge is Professor Emeritus and Endowment Fellow at the University of Edinburgh where he was Professor of Logic and Metaphysics from 1979 until 1989; previously he was a Lecturer and then Reader in Philosophy at the University of Sussex. His philosophical books are *Facts, Words and Beliefs* (1968); *Santayana: An Examination of his Philosophy* (1974); *The Vindication of Absolute Idealism* (1983); *Theories of Existence* (1984); *The Rational Foundations of Ethics* (1987). His book on F. H. Bradley and William James should be appearing fairly soon.

Michael Weston teaches philosophy at the University of Essex. He is the author of *Morality and the Self* and of articles on ethics, philosophy of religion and philosophy of literature.

Paul Williams is Lecturer in Indo-Tibetan Studies at the University of Bristol. He read Philosophy and Religion in the University of Sussex's School of African and Asian Studies and completed a D.Phil. in Madhyamaka Buddhist Philosophy at the Oriental Institute, Oxford. He is the author of many articles and reviews, mainly on Madhyamaka thought, and one book: *Mahāyāna Buddhism: The Doctrinal Foundations* (1989). He is currently European Secretary of the International Association of Buddhist Studies.

Rowan Williams was born in Wales, studied Theology at Cambridge, and did research at Oxford in Russian religious thought; he is the author of several books on the history of Christian theology and spirituality, most recently (1991) a study of Teresa of Avila, and of some shorter studies on the frontiers of theology and philosophy. Professor of Divinity at Oxford, 1986–92; (Anglican) Bishop of Monmouth 1992.

Index

Index

Index

God's justice, 223
Goethe, 68
Goldstein, Valerie Saiving, 70
good, the, 60
  desire for, 222
googols, 74
Gospel of Thomas, 215–217
Gospel of Truth, 215
Great perfection, 207
Green, T. H., 105
Gregory of Nyssa, 62–64, 66–67, 70, 222*n*
Gregory Palamas, 89–91, 98–101

Haldane, J., 41*n*, 76
heart, 91, 100
Heath, Peter, 193
Heidegger, 12, 18
  Basic Problems of Phenomenology, 19
  Basic Writings, 19
  Being and Time, 18–19
  History of the Concept of Time, 22
  Piety of Thinking, 19, 28
Hegel, 9, 11, 34, 38, 54–55, 73, 160, 211
Henle, P., 199*n*
Hepburn, R. W., 145, 148
Herbert of Cherbury, 164
Hesiod, 160
Hick, John, 53–56
Hinduism, 56
Hölderlin, 53
hope, 147
Hopkins, Gerard Manley, 39
Hopkins, J., 220*n*
Houston, G., 191, 191*n*, 194, 194*n*
Hume, 84, 164
Hunt, Holman, 49
hypostatization, 192, 197

'I', 213–214
  Cartesian, 92
'I ought', 239
'I will', 239
Ibn Bhājja, 182
Ibn Rushd (Averroes), 178
images, 129, 132, 137
  revealed, 131

imagination, 127–129, 138
  and metaphorical discourse, 142
  and the self, 130
  religious, 127–143 *passim*
imago dei, 219
imitation, 163 *et seq.*
immanence, 9–10, 15
immutability, 202
imperatives, 239
  hypothetical, 237
'impressions move the soul', 230
'individualism', 93–94
  Cartesian, 102
ineffability, 199*n*
innocence, 'context of', 239
intentionality, 69, 229, 231
Islam, 7, 56, 177–186 *passim*
ittiḥād, 182
ittiṣāl, 182
iustitia, 222

Jackson, D. P., 197*n*
James, William, 105–124 *passim*
Jaspers, Karl, 54, 133
Jesus, 53, 107, 133, 134, 140, 161, 164, 169–173, 215–217
Jo nang tradition, 208*n*
Jung, 137
Jungianism, 214*n*

Kafka, 212
kalām, 177
Kamalaśīla, 192 *et seq.*, 201, 203, 205
Kant, 93, 130, 133, 134, 147–148, 160, 239
Karma pa, Eighth, 206*n*
Karmay, S. G., 207
Karnataka, 189*n*
Kāśyaparivārta, 194
Keats, 137
Kenny, Anthony, 93*n*
Kerr, Fergus, 91–92, 101
Kierkegaard, 5, 7, 9–29 *passim*, 153, 226, 244
  Concluding Unscientific Post-script, 14–17, 22, 24–27
  Either/Or, 24
  Fear and Trembling, 27
  Journals and Papers, 14, 17, 22, 27
  Philosophical Fragments, 20–25

Index

Index

spiritual,
  life, the, 3, 4, 59, 61, 64, 69, 89,
    229–246 *passim*
  tradition, Christian, 225
Sprigge, T. L. S., 150, 242
Stendhal, 168
Stoicism, 7, 35–37, 76, 81, 156
Strawson, P. F., 128
subject, the, 213–214
sublime, the, 130
Sufism, 181
śūnyatā 53, 57, 201, 206, 216
supernaturalism, 105–125
Sykes, Stephen, 162
symbols, 131, 137
  ethical, 237

ta onta, 157
Tao, the, 53
taqlīd, 183
Taoism, 192*n*
Taylor, Charles, 59–61, 64–65, 67,
  71, 92
  Sources of the Self, 59–61
temperance, 229
theism, orthodox, 119
theologia, 146
theologians, (the), 71, 127, 179–180
  Christian, 135–136
  modern, 173
theological inquiry, 71
theology, 5, 6, 71, 138, 184
  Christian, 162
  liberation, 163, 173–174
  Muslim, 181
  natural, 1, 7, 161
  'negative', of Pseudo-Dionysius,
    99
  philosophical, 135
  western, 143
Tibet, 189
  Chinese takeover, 189*n*
Titian, 49
trance, 182
transcendence, 129, 153
'transcendent', 145
transcendental arguments, 129, 134
truth, 75 *et seq.*
  pragmatic conception of, 117
  ultimate, 201

as mental referent, 197
Tsong kha pa, 196*n*, 197, 200–202
Tsultrim Gyamtso Rinpoche, 208*n*
Tulku Thondup Rinpoche, 207*n*

ultimate, the, 196, 202–203, 206–207
  as essence of things, 200
unenlightenment, 194–195, 199, 203
'unselfing', 59, 70

Vaidya, P. L., 196
Van Gogh, 50
Vermeer, 42
Vesey, Godfrey, 222*n*
via eminentiae, 149
via negativa, 149
'victimage mechanism', 165, 169, 172
Vico, 164
vikalpa, 191
violence, 163
vipassanā, 229
virginity, 62
Vishnu, 53
Viṣṇu Purāṇa, 190*n*
Voltaire, 166–167, 169
von Stael-Holstein, 194*n*

Waismann, Friedrich, 142
Wallace, A. R., 110
Wallace, M., 162, 173, 174
Walsh, Kilian, 218*n*
Ward, Mrs Humphrey, 105
Ware, Kallistos, 99
Weil, Simone, 59, 229
Wilden, A., 213*n*
whirling, 182
will, 97
  to believe, 117 *et seq.*
William of St Thierry, 95
Williams, B., 130
Williams, H. A., 5–6
Williams, P., 189*n*, 191*n*, 192*n*,
  197*n*, 200*n*, 201*n*, 206*n*, 208*n*
Williams, R., 221*n*, 222*n*
Wittgenstein, 2, 101, 139
wisdom, 31
  analytical, 195
  unchangeable, 64, 67
women, 62, 66
wu-wei, 192*n*